F. P. J. Rimrott

Introductory Orbit Dynamics

FUNDAMENTALS AND ADVANCES
IN THE ENGINEERING SCIENCES

GRUNDLAGEN UND FORTSCHRITTE
DER INGENIEURWISSENSCHAFTEN

edited by/herausgegeben von

Prof. Dr.-Ing. Wilfried B. Krätzig, Ruhruniversität Bochum
Prof. Dr.-Ing. em. Theodor Lehmann, Ruhruniversität Bochum
Prof. Dr.-Ing. Oskar Mahrenholtz, TU Hamburg

Konvektiver Impuls-, Wärme- und Stoffaustausch,
von Michael Jischa

Einführung in Theorie und Praxis der Zeitreihen- und Modalanalyse,
von Hans G. Natke

Mechanik der Flächentragwerke,
von Yavuz Basar und Wilfried B. Krätzig

Introductory Orbit Dynamics

von Fred P. J. Rimrott

Manuscripts of abstracts should be submitted to one of the editors or to Vieweg Publishing,
Faulbrunnenstraße 13, D-6200 Wiesbaden, FRG.

Fred P. J. Rimrott

Introductory Orbit Dynamics

Friedr. Vieweg & Sohn   Braunschweig / Wiesbaden

CIP-Titelaufnahme der Deutschen Bibliothek

Rimrott, Fred P. J.:
Introductory orbit dynamics/Fred P. J. Rimrott. –
Braunschweig; Wiesbaden: Vieweg, 1989
   (Fundamentals and advances in the engineering
   sciences)

ISBN 978-3-322-90339-6   ISBN 978-3-322-90338-9 (eBook)
DOI 10.1007/978-3-322-90338-9

Fred P. J. Rimrott
Department of Mechanical Engineering
University of Toronto
Toronto, Ontario
Canada M5S 1A4

Vieweg is a subsidiary company of the Bertelsmann Publishing Group.

CONTENTS

Preface

The study of the dynamics of satellites has a unique fascination for student and lecturer alike. It is not only a logical subject explainable by the few basic principles of mechanics, but has contributed so extensibly to the formulation of mechanics in the first place, and is still continuing to do so.

With the launching of Sputnik I on October 4, 1957, engineers have entered the field for good, and the study of the dynamics of spacecraft is taking its rightful place as a subject within engineering mechanics.

The primary purpose of the present text is to acquaint engineering students with the fundamentals of spacecraft orbit dynamics. The text is intended for senior undergraduate or for graduate students, as well as for engineers in the various branches of the aerospace industry.

Students using the text are expected to know the rudiments of astronomy and to have an adequate command of elementary dynamics, of differential and integral calculus, and of vector and matrix algebra. Vectors and tensors appear in matrix form, since the matrix formulation is not only well suited for computer programming, but also because it affords a quick and intelligible assessment of the problem situation, so essential for engineering practice.

Spacecraft dynamics problems are nowadays solved by computer, a circumstance which demands from the engineer a thorough understanding of the underlying fundamentals, in order to enable him to formulate such problems for efficient computation, to understand the theory behind the computations, to validate computer outputs, and to arrive quickly at approximate results.

The book avoids overly elaborate formulations. Since it is written for engineering, the text assumes that astronomical data are *known* quantities, to be used in the design of spacecraft. In this respect, the treatment of the subject matter differs fundamentally from that in Physics or Astronomy, where efforts are typically directed towards finding and establishing such data.

The text deals with orbit dynamics and is devoted to the point satellite, first in the field of a point master, and later in the field of a master of arbitrary shape. There is a thorough discussion of the Kepler orbits. Subsequently, the insertion of a satellite into its orbit is treated, and then there follows a discussion of orbital variations for a point satellite in the field of a spheroid master.

The concept of a point satellite is also used to discuss satellite behaviour when subjected to two masters, and Hill's curves, the libration points, and Roche's limit are derived.

The derivation of each significant formula is followed by the discussion of a practical sample problem, in order to acquaint the student with typical situations, typical results, and typical numerical values. For the same purpose, there are numerous problems following each chapter. The type of problems chosen covers the whole range from programmed learning type problems to engineering applications.

The most important engineering and astronomical data, the nomenclature employed, and the answers to the problems are compiled in appendices.

The conventional practice of citing detailed references to the literature has not been followed in this text because of its introductory nature. Chapters end with a list of publications under the heading of Suggested Reading for those interested in tracing work back to its sources, or for further study of the subject matter.

The prime has been used widely in the text to designate quantities of a nature similar to that of the unprimed quantity. For instance, coherence in the treatment of conic sections has been achieved by using a' for the semi-major axis of the hyperbola, in contradistinction to the semi-major axis a of the ellipse, with $a' = -a$.

Each chapter is essentially self-contained, allowing the teacher and the reader flexibility of instruction without loss of coherence.

The author is aware of the book's numerous deficiencies and imperfections, and of its falling short of the degree of excellence that might be achieved. He is left to trust in the indulgence of those for whose benefit it is intended.

Historical Background*

When François-Marie Arouet (1694-1778), who called himself Voltaire, at the time (1750-1753) academician at Frederick the Great's (1712-1786) Prussian Academy of Science in Berlin, had a disagreement with Pierre Louis Moreau de Maupertuis** (1698-1759), the Academy's President (1741-1756), he ridiculed him in his *Le Diatribe du Docteur Akakia* (1753), which lists all the president's harebrained schemes: Among them Maupertuis' assertion, that it would be possible, one day, to travel to the Moon.

Tsiolkovsky (1857-1935), Goddard (1882-1945) and Oberth (1894-   ), the three wise men whose vision and patient persistence has led directly and triumphantly to man's actual entry into space in our time, by the large are remembered and honoured only by their peers and successors.

Jules Verne's (1828-1905) science romances, such as *De la terre à la lune* (1865), achieved instant and lasting popularity, but when the Russian school teacher Konstantin Eduardovich Tsiolkovsky in 1896 published a serious article on communications with the inhabitants of other planets, very little notice was taken of it. That same year, Tsiolkovsky began work on his *Exploration of Cosmic Space by Means of Reaction Devices*. The Russian scientific community, however, remained totally indifferent to his theories, his writings and his aerodynamic experiments, and so did the rest of the world.

Recognition did come in the end. In 1919, Tsiolkovsky was elected to the Academy of Sciences, and two years later the Soviet government awarded him a life pension.

The endless rocket experiments of the American Dr. Robert H. Goddard (the first man to demonstrate that thrust and propulsion can perform in a vacuum and requires no air to push against) continued for many years to cause no more than tolerant amusement. It was thought hilarious that a Ph.D., a physicist with Goddard's impressive credentials, "could take seriously the possibility of travel beyond Earth".

Yet the space age may be said to have begun, not on October 4, 1957, when Sputnik I first went into orbit around the Earth, but on March 26, 1926, when - the reference books record - "the world's first flight of a liquid-propelled rocket engine took place on the Auburn, Massachusetts, farm of Goddard's Aunt Effie".

At the time - in 1919 - when he published his classic treatise *A Method of Reaching Extreme Altitudes*, Goddard was unaware of the work and the experiments of Tsiolkovsky, whose book on space exploration had appeared - and been ignored - in 1903, nor had Hermann Oberth, the third of the pioneer trio, seen Goddard's paper when he, Oberth, published his *Die Rakete zu den Planetenräumen* in 1923.

Much later, the three physicists did communicate with one another, and Oberth, now living in retirement in Germany, has always acknowledged the precedence of the other two in the development of the mathematical theory of space flight.

Tsiolkovsky was ignored, Goddard was ridiculed, and when Oberth in 1922 sought a doctor's degree from Heidelberg University, his thesis on rocket flight was rejected. Yet their concepts have become reality, and - looking at any rocket-launched spacecraft of today (or tomorrow) - we do well to ask, as does the aviator Robur in Verne's 1886 novel *Clipper of the Clouds*, "Without the groping experiments of his predecessors, could an engineer ever have conceived so perfect an apparatus?"

Toronto, Ontario, Canada   *F.P.J. Rimrott*

---

*adapted from Spar News, IX, 4, December 1977.

**Maupertuis' fame by the way rests solidly on the pronouncement of the Principle of Least Action, and upon proving by geodesic measurements that the Earth is flattened at the poles.

1

The Simplified Two-Body Problem

Satellites[*] move within the force field of their masters. Artificial satellites have a mass m which is so small compared to the mass M of their master, that we may simplify the problem by setting

$$m << M$$

This permits us to assume on the one hand that the satellite's motion is dictated by the master, and that, on the other hand, the master's motion is not influenced by the satellite. For convenience, we can now assume that the master is fixed in space and that the satellite moves around the master. We place the origin of the coordinate system to be used to describe the motion of the satellite at the mass centre of the master mass M. Additional conciseness in the theory is achieved by assuming that the master mass is concentrated in a point, which then represents a fixed *centre* towards which the satellite is attracted and about which the satellite moves, circumstances that have given rise to the term *central force motion* for that field of study. If the satellite mass is also assumed to be concentrated in a point, then relations that were first stated by Kepler[**] (in 1609 and 1619) are obtained, and terms such as *Kepler orbit* and *Kepler force* will subsequently be used to imply central force motion of point satellites.

The satellite m obeys Newton's second law[***]

$$\mathbf{K} = m\ddot{\mathbf{r}} \tag{1.1}$$

where $\mathbf{K}$ is the Kepler force, i.e. the attraction force acting on the satellite, and $\ddot{\mathbf{r}}$ is the absolute acceleration of the satellite.

1.1 Position, Velocity and Acceleration

In order to describe the motion of the satellite, we employ a *cylindrical coordinate* system with the unit vectors (Figure 1.1).

$$
\begin{aligned}
\mathbf{e}_r &\quad \text{in } \textit{radial} \text{ direction} \\
\mathbf{e}_\theta &\quad \text{in } \textit{transverse} \text{ direction} \\
\mathbf{e}_z &\quad \text{in } \textit{normal} \text{ direction}
\end{aligned}
$$

The position $\mathbf{r}$ of the point satellite m (Figure 1.2) is

$$\mathbf{r} = \begin{bmatrix} \mathbf{e}_r & \mathbf{e}_\theta & \mathbf{e}_z \end{bmatrix} \begin{bmatrix} r \\ 0 \\ 0 \end{bmatrix} \tag{1.2}$$

The first derivative of the position vector $\mathbf{r}$ with respect to time is the *velocity* $\dot{\mathbf{r}}$

$$\dot{\mathbf{r}} = \begin{bmatrix} \mathbf{e}_r & \mathbf{e}_\theta & \mathbf{e}_z \end{bmatrix} \begin{bmatrix} \dot{r} \\ 0 \\ 0 \end{bmatrix} + \begin{bmatrix} \dot{\mathbf{e}}_r & \dot{\mathbf{e}}_\theta & \dot{\mathbf{e}}_z \end{bmatrix} \begin{bmatrix} r \\ 0 \\ 0 \end{bmatrix} \tag{1.3}$$

---

[*] From the Latin, *satelles, itis = body guard*.
[**] Johannes Kepler (1571-1630), German astronomer.
[***] Isaac Newton (1642-1727), English physicist.

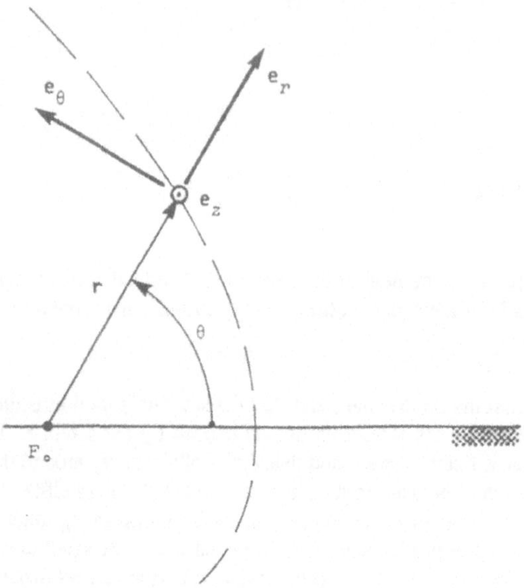

FIGURE 1.1. Standard basis vectors for cylindrical coordinate system.

We can write this equation also in the following form

$$\dot{\mathbf{r}} = \overset{\circ}{\mathbf{r}} + \boldsymbol{\omega} \times \mathbf{r} \tag{1.4}$$

($\overset{\circ}{\mathbf{r}}$ is pronounced "r circle"). The first term on the right hand side of equation (1.4) is the *relative velocity*

$$\overset{\circ}{\mathbf{r}} = \begin{bmatrix} \mathbf{e}_r\ \mathbf{e}_\theta\ \mathbf{e}_z \end{bmatrix} \begin{bmatrix} \dot{r} \\ 0 \\ 0 \end{bmatrix} \tag{1.5}$$

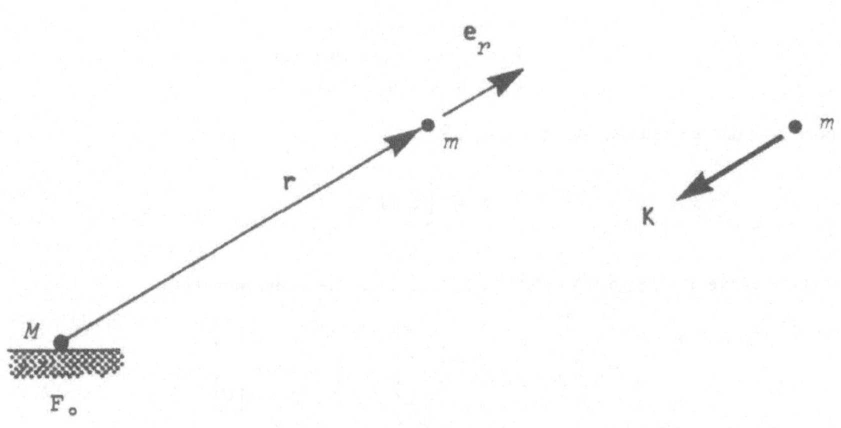

FIGURE 1.2. Attraction of a satellite m by its master M.

The second term is the *carrier point velocity* (or *guide velocity* or *lead velocity*) of the carrier point coinciding momentarily with m, the carrier in this instance being the $F_o r\theta z$ coordinate system,

$$\omega \times r = \begin{bmatrix} \dot{e}_r & \dot{e}_\theta & \dot{e}_z \end{bmatrix} \begin{bmatrix} r \\ 0 \\ 0 \end{bmatrix} \tag{1.6}$$

where ω is the angular velocity of the carrier. In the present case, the rotation of the coordinate system takes place about the z-axis. Thus the angular velocity of the coordinate system, i.e. of the carrier, is

$$\omega = \begin{bmatrix} e_r & e_\theta & e_z \end{bmatrix} \begin{bmatrix} 0 \\ 0 \\ \dot{\theta} \end{bmatrix} \tag{1.7}$$

For the cross product, we obtain

$$\omega \times r = \begin{vmatrix} e_r & e_\theta & e_z \\ 0 & 0 & \dot{\theta} \\ r & 0 & 0 \end{vmatrix} = \begin{bmatrix} e_r & e_\theta & e_z \end{bmatrix} \begin{bmatrix} 0 \\ r\dot{\theta} \\ 0 \end{bmatrix} \tag{1.8}$$

For the velocity v of the satellite m as given by equation (1.3), we eventually obtain, by adding equations (1.5) and (1.8),

$$v - \dot{r} = \begin{bmatrix} e_r & c_\theta & c_z \end{bmatrix} \begin{bmatrix} \dot{r} \\ r\dot{\theta} \\ 0 \end{bmatrix} \tag{1.9}$$

We observe that the velocity of the satellite may be resolved into a *radial* and a *transverse* component.

The second derivative of the position vector r with respect to time is the *acceleration* $\ddot{r}$. Applying the mathematical rule expressed by equation (1.4) to equation (1.4) twice, we obtain

$$\ddot{r} = (\dot{r})^\cdot = \overset{\circ}{\dot{r}} + \omega \times \dot{r} = (\overset{\circ}{r} + \omega \times r)^\circ + \omega \times \overset{\circ}{r} + \omega \times (\omega \times r) \tag{1.10}$$

giving eventually

$$\ddot{r} = \overset{\circ\circ}{r} + 2\omega \times \overset{\circ}{r} + \overset{\circ}{\omega} \times r + \omega \times (\omega \times r) \tag{1.11}$$

where $\overset{\circ\circ}{r}$ is the *relative acceleration*, $2\omega \times \overset{\circ}{r}$ is the *Coriolis supplementary acceleration*, and $\overset{\circ}{\omega} \times r + \omega \times (\omega \times r)$ is the *carrier point acceleration*.

For the present case, we have ($\overset{\circ\circ}{r}$ is pronounced "r double circle")

$$\overset{\circ\circ}{r} = \begin{bmatrix} e_r & e_\theta & e_z \end{bmatrix} \begin{bmatrix} \ddot{r} \\ 0 \\ 0 \end{bmatrix}$$

$$\overset{\circ}{\omega} = \begin{bmatrix} e_r & e_\theta & e_z \end{bmatrix} \begin{bmatrix} 0 \\ 0 \\ \ddot{\theta} \end{bmatrix}$$

$$\overset{\circ}{\omega} \times r = \begin{vmatrix} e_r & e_\theta & e_z \\ 0 & 0 & \ddot{\theta} \\ r & 0 & 0 \end{vmatrix} = \begin{bmatrix} e_r & e_\theta & e_z \end{bmatrix} \begin{bmatrix} 0 \\ r\ddot{\theta} \\ 0 \end{bmatrix}$$

$$2\omega \times \overset{\circ}{r} = 2\begin{vmatrix} e_r & e_\theta & e_z \\ 0 & 0 & \dot{\theta} \\ \dot{r} & 0 & 0 \end{vmatrix} = \begin{bmatrix} e_r & e_\theta & e_z \end{bmatrix} \begin{bmatrix} 0 \\ 2\dot{r}\dot{\theta} \\ 0 \end{bmatrix}$$

$$\omega \times (\omega \times r) = \begin{vmatrix} e_r & e_\theta & e_z \\ 0 & 0 & \dot\theta \\ 0 & r\dot\theta & 0 \end{vmatrix} = \begin{bmatrix} e_r, & e_\theta & e_z \end{bmatrix} \begin{bmatrix} -r\dot\theta^2 \\ 0 \\ 0 \end{bmatrix}$$

After adding all terms, the final expression for the acceleration **a** of the satellite becomes

$$\mathbf{a} = \ddot{\mathbf{r}} = \begin{bmatrix} e_r, & e_\theta & e_z \end{bmatrix} \begin{bmatrix} \ddot{r} - r\dot\theta^2 \\ r\ddot\theta + 2\dot{r}\dot\theta \\ 0 \end{bmatrix} \tag{1.12}$$

The acceleration thus consists also of a radial and a transverse component.

1.2 The Attraction Force

Point satellite m is attracted by the point master M by a force

$$\mathbf{K} = -\frac{GMm}{r^2}\mathbf{e}_r \tag{1.13}$$

Equation (1.13) represents the *universal gravitational law* (also called *inverse square law*) first enunciated by Newton. The attraction force **K** is directed towards the point master; the quantity r is the distance between point master and point satellite (Figure 1.2). The quantity G is the *universal gravitational constant*.

$$G = 6.67 \cdot 10^{-11} \frac{m^3}{kg\ s^2} \tag{1.14}$$

where s is a second of *solar time* (Appendix A).

The product GM is often combined into the *gravitational parameter*, μ, a quantity characteristic for each master.

$$\mu = GM \tag{1.15}$$

For the Earth

$$\mu_E = 398\ 601.19\ km^3/s^2 \tag{1.16}$$

for the Moon

$$\mu_M = 4902.73\ km^3/s^2 \tag{1.17}$$

for Mars

$$\mu_{Mars} = 42\ 828.3\ km^3/s^2 \tag{1.18}$$

Combining equations (1.13) and (1.15), we can express the attraction force as

$$\mathbf{K} = \begin{bmatrix} e_r, & e_\theta & e_z \end{bmatrix} \begin{bmatrix} -\dfrac{\mu m}{r^2} \\ 0 \\ 0 \end{bmatrix} \tag{1.19}$$

1.3 Shape of Path

In order to ascertain the shape of the path on which a satellite orbits its master, we apply Newton's second law

$$\mathbf{K} = m\ddot{\mathbf{r}} \tag{1.20}$$

Referring to equations (1.9) and (1.19), one can write

$$\begin{bmatrix} \mathbf{e}_r\ \mathbf{e}_\theta\ \mathbf{e}_z \end{bmatrix} \begin{bmatrix} -\dfrac{\mu m}{r^2} \\ 0 \\ 0 \end{bmatrix} = \begin{bmatrix} \mathbf{e}_r\ \mathbf{e}_\theta\ \mathbf{e}_z \end{bmatrix} \begin{bmatrix} m(\ddot{r} - r\dot{\theta}^2) \\ m(r\ddot{\theta} + 2\dot{r}\dot{\theta}) \\ 0 \end{bmatrix} \tag{1.21}$$

The mass m of the satellite can be cancelled out and we consequently conclude that *the mass of a satellite has no influence on its acceleration.*

Equation (1.21) contains a system of two coupled nonlinear differential equations,

$$\ddot{r} - r\dot{\theta}^2 - = -\frac{\mu}{r^2} \tag{1.22a}$$

$$r\ddot{\theta} + 2\dot{r}\dot{\theta} = 0 \tag{1.22b}$$

Explicit solutions $r = r(t)$ and $\theta = \theta(t)$ in closed form cannot be found. There are, however, two integrable combinations which satisfy the differential equations, namely

$$r^2\dot{\theta} = h \tag{1.23}$$

$$r(1 + \varepsilon \cos(\theta + A_1)) = \frac{h^2}{\mu} \tag{1.24}$$

The quantities h, ε and A_1 are integration constants. The quantity h is the *specific angular momentum* of the satellite with respect to the master. It is related to the angular momentum H of a point satellite about the occupied focus F_o (Figure 1.3) by

$$h = \frac{H}{m} \tag{1.25}$$

The quantity ε is the *(numerical) eccentricity* of the Kepler orbit. The quantity A_1 represents a phase angle whose magnitude depends on the choice of base line from which the polar angle θ is measured. One can choose $A_1 = 0$, which means that the polar angle is measured from the *line of apsides* from the branch that passes through the orbit's *periapsis* PE. If the polar angle is furthermore measured in the direction of motion of the satellite, then it becomes the *true anomaly* (Figure 1.3).

Solving for r and setting $A_1 = 0$, equation (1.24) assumes the following appearance

$$r = \frac{\dfrac{h^2}{\mu}}{1 + \varepsilon \cos\theta} \tag{1.26}$$

i.e. the appearance of the polar equation of a conic section, which is usually written

$$r = \frac{p}{1 + \varepsilon \cos\theta} \tag{1.27}$$

A comparison of equations (1.26) and (1.27) indicates that *semi-parameter p*, angular momentum h, and gravitational parameter μ, are related by

$$p = \frac{h^2}{\mu}$$

or

$$h = \sqrt{\mu p} \tag{1.28}$$

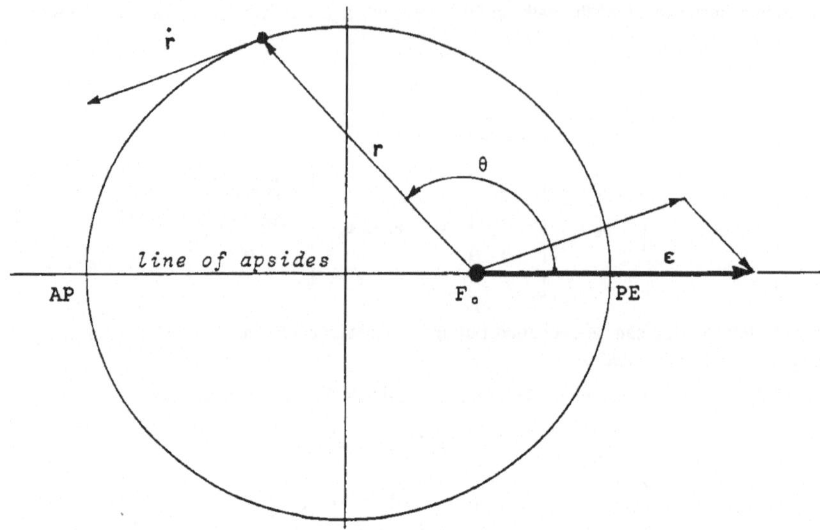

FIGURE 1.3. The eccentricity vector resolved into components along $\dot{r}$ and r.

The point PE on an orbit (Figure 1.3) closest to the occupied focus F_o is the *periapsis* (or *pericentre*). For orbits around the Earth the *periapsis* is also called the *perigee*, for orbits around the Sun the *perihelion*, for orbits around the Moon the *perilun* (or *perisel*), etc.

The point AP on an elliptic orbit, furthest from the occupied focus F_o is the *apoapsis* (or *apocentre*). For orbits around the Earth the *apoapsis* is also called the *apogee*, for orbits around the Sun the *aphelion*, for orbits around the Moon the *apolun* (or *aposel*), etc.

1.4   The Eccentricity Vector

Equation (1.21) can also be integrated in vector form. We begin with

$$\ddot{\mathbf{r}} = -\frac{\mu}{r^3}\mathbf{r}$$

and cross-multiply with the specific angular momentum vector $\mathbf{h}$

$$\mathbf{h}\times\ddot{\mathbf{r}} = \frac{\mu}{r^3}(\mathbf{r}\times\mathbf{h})$$

Since $d(\mathbf{h}\times\dot{\mathbf{r}})/dt = \mathbf{h}\times\ddot{\mathbf{r}}$ because $\mathbf{h} = $ constant, since $\mathbf{h} = \mathbf{r}\times\dot{\mathbf{r}}$, and since $\mathbf{r}\times(\mathbf{r}\times\dot{\mathbf{r}}) = (\mathbf{r}\cdot\dot{\mathbf{r}})\mathbf{r} - (\mathbf{r}\cdot\mathbf{r})\dot{\mathbf{r}}$, we obtain

$$\frac{d}{dt}(\mathbf{h}\times\dot{\mathbf{r}}) = \frac{\mu}{r^3}((\mathbf{r}\cdot\dot{\mathbf{r}})\mathbf{r} - (\mathbf{r}\cdot\mathbf{r})\dot{\mathbf{r}})$$

As $\mathbf{r}\cdot\mathbf{r} = r^2$ and $\mathbf{r}\cdot\dot{\mathbf{r}} = r\dot{r}$, one can write

$$\frac{d}{dt}(\mathbf{h}\times\dot{\mathbf{r}}) = \frac{\mu\dot{r}}{r^2}\mathbf{r} - \frac{\mu}{r}\dot{\mathbf{r}}$$

Since $d(\mathbf{r}/r)/dt = \dot{\mathbf{r}}/r - \dot{r}\mathbf{r}/r^2$, we can write

$$\frac{d}{dt}(\mathbf{h}\times\dot{\mathbf{r}}) = -\mu\frac{d}{dt}\left[\frac{\mathbf{r}}{r}\right]$$

which may be integrated to give

$$\dot{\mathbf{r}} \times \mathbf{h} = \mu \frac{\mathbf{r}}{r} + \mathbf{C}$$

It can be shown that the integration constant $\mathbf{C}$ is a vector pointing towards periapsis. The vector $\boldsymbol{\varepsilon} = \mathbf{C}/\mu$ is called the *eccentricity vector* and given by

$$\boldsymbol{\varepsilon} = \frac{1}{\mu}(\dot{\mathbf{r}} \times \mathbf{h}) - \frac{\mathbf{r}}{r} \tag{1.29}$$

It points from the occupied focus F_o towards the periapsis PE (Figure 1.3). Its magnitude is equal to the orbit's eccentricity.

The specific angular momentum $\mathbf{h}$ can be eliminated by recalling that $\mathbf{h} = \mathbf{r} \times \dot{\mathbf{r}}$. Observing that $\dot{\mathbf{r}} \cdot \dot{\mathbf{r}} = v^2$ and that $\dot{\mathbf{r}} \cdot \mathbf{r} = r\dot{r}$, one obtains

$$\boldsymbol{\varepsilon} = \left[\frac{v^2}{\mu} - \frac{1}{r}\right]\mathbf{r} - \frac{r\dot{r}}{\mu}\dot{\mathbf{r}} \tag{1.30}$$

or

$$\boldsymbol{\varepsilon} = \left[1 - \frac{r}{a}\right]\frac{\mathbf{r}}{r} - \frac{r\dot{r}v}{\mu}\frac{\dot{\mathbf{r}}}{v} \tag{1.31}$$

where both $\mathbf{r}/r$ and $\dot{\mathbf{r}}/v$ are unit vectors.

As an example, consider a satellite orbiting the Earth with $r_{PE} = 7000$ km and $r_{AP} = 10000$ km, at a true anomaly of $120°$. For the solution, we calculate

$$p = \frac{2\,r_{PE}r_{AP}}{r_{AP} + r_{PE}} = \frac{2(7000)\,10\,000}{10\,000 + 7000} = 8235\ \text{km}$$

$$\varepsilon = \frac{r_{AP} - r_{PE}}{r_{AP} + r_{PE}} = \frac{10\,000 - 7000}{10\,000 + 7000} = 0.176\,47$$

$$r = \frac{p}{1 + \varepsilon\cos\theta} = \frac{8235}{1 + 0.176\cos 120°} = 9032\ \text{km}$$

$$a = \frac{1}{2}\left[r_{AP} + r_{PE}\right] = 8500\ \text{km}$$

$$v^2 = \mu\left(\frac{2}{r} - \frac{1}{a}\right) = 41.370\ \text{km}^2/\text{s}^2$$

$$\dot{r} = \sqrt{\frac{\mu}{p}}\ \varepsilon\sin\theta = 1.063\ \text{km/s}$$

to obtain eventually (Figure 1.3)

$$\boldsymbol{\varepsilon} = -0.0\,626\frac{\mathbf{r}}{r} - 0.155\frac{\dot{\mathbf{r}}}{v}$$

Suggested Reading

1. Geyling, F.T.; H.R. Westerman. *Introduction to Orbital Mechanics*, Addison-Wesley, 1971, 349 pp.

2. Kaplan, M.H. *Modern Spacecraft Dynamics and Control*, John Wiley and Sons, 1976, 415 pp.

3. Oberth, H. *Wege zur Raumschiffahrt*, Kriterion-Verlag, Bukarest, 1974, 414 pp.

4.   Roy, A.E. *The Foundations of Astrodynamics*, Macmillan, 1965, 385 pp.

5.   Thomson, W.T. *Introduction to Space Dynamics*, John Wiley and Sons, 1963, 317 pp.

Problems

1.1   The angular momentum (per unit mass) for a point satellite is given by

$$h = \sqrt{\mu a(1 - \varepsilon^2)}$$

(a) Determine h for a satellite whose $a = 8135$ km and $\varepsilon = 0.17$; and (b) determine h for the Moon, whose $a = 384\,400$ km and $\varepsilon = 0.0549$.

1.2   Sketch six ellipses. All have the same major axis. Their eccentricities, however, are

$$\varepsilon = 0, 0.2, 0.4, 0.6, 0.8, 1.0$$

1.3   The attraction force between two bodies is

$$F = \frac{GMm}{r^2}$$

Who attracts the Moon more strongly: the Sun ($M_S = 1.989 \cdot 10^{30}$ kg, $r_S = 1.5 \cdot 10^8$ km) or the Earth ($M_E = 5.976 \cdot 10^{24}$ kg, $r_E = 384\,400$ km)?

1.4   Show that

$$\left[\dot{e}_x\, \dot{e}_y\, \dot{e}_z \right] \begin{bmatrix} x \\ y \\ z \end{bmatrix} = \boldsymbol{\omega} \times \mathbf{r}$$

if

$$\mathbf{r} = \left[e_x\, e_y\, e_z \right] \begin{bmatrix} x \\ y \\ z \end{bmatrix} \qquad \text{with} \quad x = \text{constant}, y = \text{constant}, z = \text{constant}$$

and

$$\boldsymbol{\omega} = \left[e_x\, e_y\, e_z \right] \begin{bmatrix} \omega_x \\ \omega_y \\ \omega_z \end{bmatrix}$$

1.5   Show that

$$r^2 \dot{\theta} = h$$

$$r(1 + \varepsilon \cos(\theta + A_1)) = \frac{h^2}{\mu}$$

are integrable combinations of the differential equation system

$$\ddot{r} - r\dot{\theta}^2 = -\frac{\mu}{r^2}$$

$$r\ddot{\theta} + 2\dot{r}\dot{\theta} = 0$$

1.6   How is the eccentricity vector

$$\boldsymbol{\varepsilon} \;=\; \left[1 - \frac{r}{a}\right]\frac{\mathbf{r}}{r} \;-\; \frac{r\dot{r}v}{\mu}\frac{\dot{\mathbf{r}}}{v}$$

(where $|\dot{\mathbf{r}}| \;=\; v$) related to the numerical eccentricity ε? (*Hint*: Write $\mathbf{r}\cdot\boldsymbol{\varepsilon} \;=\; r\varepsilon\cos\theta$)

1.7   Show that the orbit eccentricity vector $\boldsymbol{\varepsilon}$ points from the occupied focus F_o to the orbit's periapsis. (*Hint*: Form $\boldsymbol{\varepsilon}\times\mathbf{r}$ and show that its magnitude equals $\varepsilon r\sin\theta$, and that its direction is perpendicular to the orbital plane.) Make a sketch.

1.8   A coordinate is called ignorable if it is missing from the Lagrangian $L \;=\; T \;-\; U$, with $T = mv^2/2$ and $U = -\mu m/r$. Determine the Lagrangian of the Kepler problem, and (a) show that θ is an ignorable coordinate. Use (b) the Lagrangian in the Lagrange equation

$$\frac{d}{dt}\frac{\partial L}{\partial \dot{q}_i} \;-\; \frac{\partial L}{\partial q_i} \;=\; 0$$

with $q_1 = r$ and $q_2 = \theta$, to obtain the differential equations of motion.

1.9   Determine the elements of the matrix $[\omega]$ in the relationship

$$\boldsymbol{\omega}\times\mathbf{r} \;=\; \{e\}^T[\omega]\{r\}$$

with

$$\{e\}^T \;=\; \left[e_x\,e_y\,e_z\right]$$

and

$$\{r\} \;=\; \begin{bmatrix} x \\ y \\ z \end{bmatrix}$$

1.10   When using a spherical coordinate system, the position vector to a point P is

$$\mathbf{r} \;=\; \left[e_r\,e_\theta\,e_\phi\right]\begin{bmatrix} r \\ 0 \\ 0 \end{bmatrix}$$

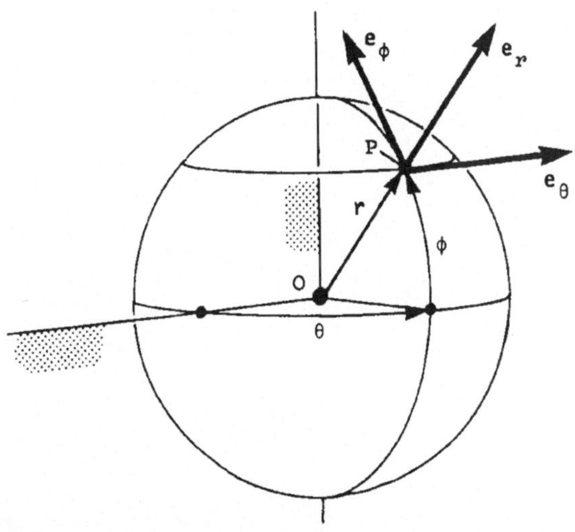

Show (a) that the velocity of point P is

$$\dot{\mathbf{r}} = \begin{bmatrix} \mathbf{e}_r\ \mathbf{e}_\theta\ \mathbf{e}_\phi \end{bmatrix} \begin{bmatrix} \dot{r} \\ r\dot{\theta}\cos\phi \\ r\dot{\phi} \end{bmatrix}$$

and (b) that the acceleration of point P is

$$\ddot{\mathbf{r}} = \begin{bmatrix} \mathbf{e}_r\ \mathbf{e}_\theta\ \mathbf{e}_\phi \end{bmatrix} \begin{bmatrix} \ddot{r} - r\dot{\phi}^2 - r\dot{\theta}^2\cos^2\phi \\ \dfrac{\cos\phi}{r}\dfrac{d}{dt}(r^2\dot{\theta}) - 2r\dot{\theta}\dot{\phi}\sin\phi \\ \dfrac{1}{r}\dfrac{d}{dt}(r^2\dot{\phi}) + r\dot{\theta}^2\sin\phi\cos\phi \end{bmatrix}$$

2

Kepler Orbits

The path of a point satellite in the field of a point master is called a *Kepler orbit*. The point master is located in one focus, called the *occupied focus* (or *prime focus*) F_o of a Kepler orbit. The second focus is unoccupied and is referred to as the *vacant focus* F_v (Figure 2.1). Kepler orbits are *conic sections*, whose equation in polar coordinates with origin at the occupied focus is

$$r = \frac{p}{1 + \varepsilon \cos\theta} \tag{2.1}$$

with p as *semi-parameter* (or *semi-latus rectum*), and ε as *(numerical) eccentricity*.

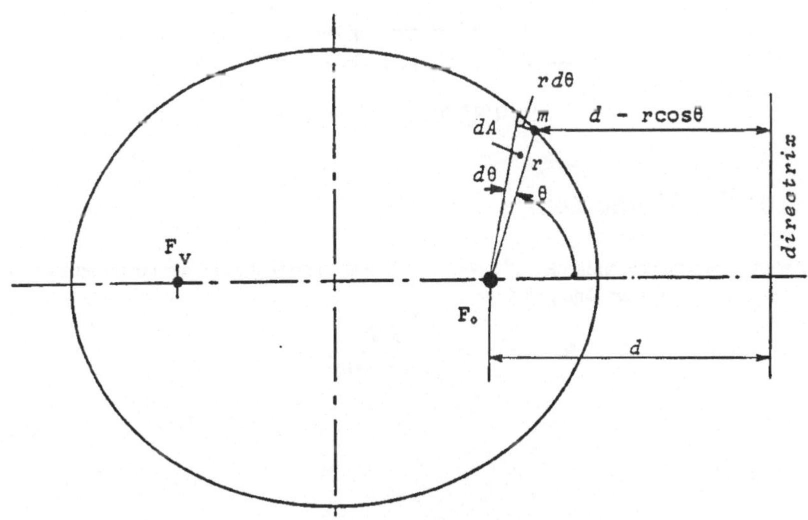

FIGURE 2.1. Area element.

A conic section is a plane curve formed by the locus of a point, m in Figure 2.1, such that the ratio ε of its distance r from a point F_o to its distance $(d - r\cos\theta)$ from a straight line, called the *directrix*, is constant.

$$\varepsilon = \frac{r}{d - r\cos\theta} = \text{constant}$$

Conic sections (Figure 2.2) include the circle, the ellipse (Figure 2.3), the parabola (Figure 2.4), and the hyperbola (Figure 2.5). Geometrical relationships for the ellipse are listed in Table 2.1, for the parabola in Table 2.2, and for the hyperbola in Table 2.3.

For a discussion of Kepler orbits for large satellites see Schiehlen and Kolbe (1969).

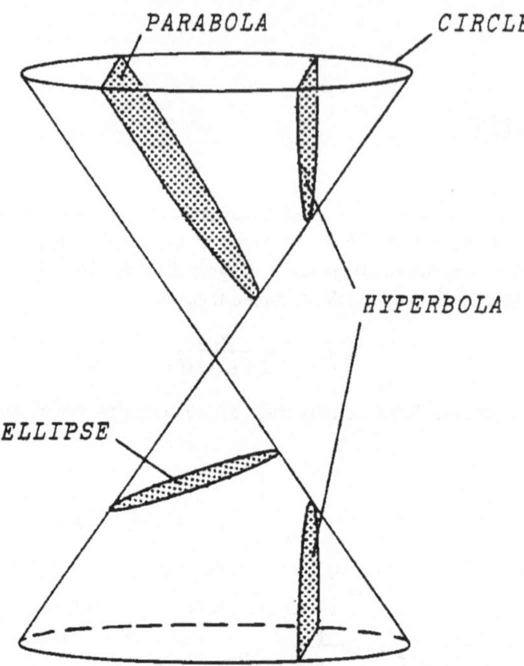

FIGURE 2.2. Conic sections.

2.1 Angular Momentum

The angular momentum h per unit mass of a point satellite with respect to its master is constant. It is called the *specific angular momentum*, and given by

$$h = r^2\dot\theta = \frac{p^2\dot\theta}{(1 + \varepsilon \cos\theta)^2} = \sqrt{\mu p} \tag{2.2}$$

2.2 Period

An inspection of Figure 2.1 indicates that the area element dA swept out by the radius r is

$$dA = \frac{1}{2} r^2 d\theta$$

From equation (2.2), we have

$$r^2 \frac{d\theta}{dt} = h$$

such that

$$dA = \frac{1}{2} h \, dt$$

For a full orbit ($\int dA = \pi ab$ for the area of an ellipse), we obtain

$$\pi ab = \frac{1}{2} h \int_o^\tau dt$$

where τ is the *Kepler period*. Solved for τ

$$\tau = \frac{2\pi ab}{h}$$

If one now considers that $h = \sqrt{\mu p} = \sqrt{\mu a(1-\varepsilon^2)}$ and that $b = a\sqrt{1-\varepsilon^2}$, then the Kepler period becomes

$$\tau = 2\pi\sqrt{\frac{a^3}{\mu}} \tag{2.3}$$

In astronomy it is customary to refer to $\sqrt{\mu/a^3}$ as the *mean angular speed n*.

$$n = \sqrt{\frac{\mu}{a^3}} \tag{2.4}$$

The (theoretical) lower limit of orbits of a satellite about a given master is a circular orbit with radius R of the master's surface. The attraction force at the master's surface is $mg = \mu m/R^2$. Entered into equation (2.3), one obtains the *Schuler period*[*]

$$\tau_s = 2\pi\sqrt{\frac{R}{g}} \tag{2.5}$$

which represents a lower bound on the orbital period of a point satellite, about a spherical master. With the Earth as master, using $R = 6371$ km and $g = 9.82027$ m/s^2 (and ignoring the presence of the atmosphere), the Schuler period is

$$84.347 \text{ min}$$

For satellites of the Moon, using $R = 1738$ km and $g = 1.623$ m/s^2, the Schuler period amounts to

$$108.364 \text{ min}$$

in spite of the smaller size of the Moon! For a detailed discussion of the significance of the Schuler period see Magnus (1978).

Synchronous satellites are satellites circling the master in its equatorial plane at a distance which causes the satellite's angular velocity to coincide with the angular velocity of the master's spin. Viewed from any point on the surface of the master, synchronous satellites thus appear to be *stationary* in the sky. The Kepler period of a synchronous satellite is equal to the spin period of the master. A synchronous satellite of the Earth has thus a Kepler period of one sidereal day, i.e.

$$\tau = 23 \text{ h } 56 \text{ min } 4.09 \text{ s} = 1436.07 \text{ min}$$

which places it at $r = 42\,164.23$ km from Earth centre, above the equator. Engineers refer to such satellites as *geostationary*, or *geosynchronous* (Figure 2.6).

---

[*]Max Schuler (1882-1972), German engineer.

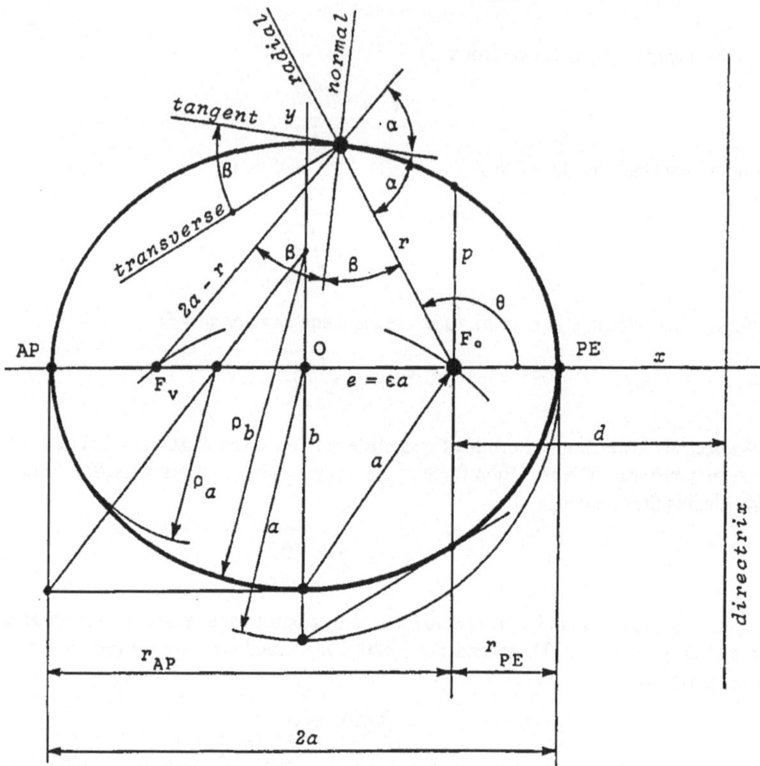

FIGURE 2.3. Ellipse.

2.3   Vis-Viva Integral

According to equation (2.1), the radius vector is

$$r = \frac{p}{1 + \varepsilon \cos\theta}$$

From equation (2.2), the transverse velocity component is

$$v_\theta = r\dot\theta = \frac{\sqrt{\mu p}}{r} \tag{2.6}$$

The radial velocity component is obtained by differentiating equation (2.1).

$$v_r = \dot r = \frac{p\varepsilon \sin\theta}{(1 + \varepsilon \cos\theta)^2}\,\dot\theta = \frac{r^2\dot\theta}{p}\,\varepsilon\sin\theta = \sqrt{\frac{\mu}{p}}\,\varepsilon\sin\theta \tag{2.7}$$

The square of the velocity equals the sum of the squares of radial and transverse components

$$v^2 = v_r^2 + v_\theta^2 = \dot r^2 + (r\dot\theta)^2 = \frac{\mu}{p}\,\varepsilon^2 \sin^2\theta + \frac{\mu p}{r^2}$$

Employing equation (2.1) once more, and combining terms, we obtain

$$v^2 = \frac{\mu}{p}\,(\varepsilon^2 - 1) + \frac{\mu}{p}\,2(1 + \varepsilon\cos\theta)$$

$$r = \frac{p}{1 + \varepsilon \cos\theta}$$

$$p = a(1 - \varepsilon^2) = \varepsilon d$$

$$a = \frac{p}{1 - \varepsilon^2}$$

$$b = \frac{p}{\sqrt{1 - \varepsilon^2}}$$

$$\frac{x^2}{a^2} + \frac{y^2}{b^2} = 1$$

$$p = \frac{b^2}{a}$$

$$\varepsilon = \sqrt{1 - \frac{b^2}{a^2}}$$

$$b = \sqrt{ap}$$

$$b = a\sqrt{1 - \varepsilon^2}$$

$$r_{PE} = \frac{p}{1 + \varepsilon} = a(1 - \varepsilon) = q$$

$$r_{AP} = \frac{p}{1 - \varepsilon} = a(1 + \varepsilon)$$

$$a = \frac{1}{2}\left[r_{AP} + r_{PE}\right]$$

$$b = \sqrt{r_{AP} r_{PE}}$$

$$p = \frac{2 r_{AP} r_{PE}}{r_{AP} + r_{PE}}$$

$$0 < \varepsilon < 1$$

$$e < a$$

$$e = \varepsilon a$$

$$a = \frac{b}{\sqrt{1 - \varepsilon^2}}$$

$$e^2 = a^2 - b^2$$

$$\varepsilon = \frac{r_{AP} - r_{PE}}{r_{AP} + r_{PE}}$$

$$\varepsilon = \frac{p}{r_{PE}} - 1$$

$$\varepsilon = 1 - \frac{p}{r_{AP}}$$

$$\tan\beta = \varepsilon \frac{r}{p} \sin\theta$$

$$\tan\beta = \frac{\varepsilon \sin\theta}{1 + \varepsilon \cos\theta}$$

$$\alpha + \beta = 90°$$

$$\rho_a = \frac{b^2}{a} = p$$

$$\rho_b = \frac{a^2}{b} = \frac{a^3}{b^3} p$$

$$\sin E = \frac{\sqrt{1 - \varepsilon^2} \sin\theta}{1 + \varepsilon \cos\theta}$$

$$M = E - \varepsilon \sin E$$

TABLE 2.1. Key equations for ellipses.

Eventually, we obtain the so-called *vis-viva integral*[*] as

$$v^2 = \mu\left[\frac{2}{r} - \frac{1 - \varepsilon^2}{p}\right] \tag{2.8}$$

A more concise formulation of the vis-viva integral can be obtained when the shape of the satellite's path is specified. For a circular orbit, for which $\varepsilon = 0$ and $r = p = a$, the vis-viva integral is

$$v^2 = \frac{\mu}{a} \tag{2.9}$$

---

[*] The designation *vis viva* (Latin for *living force*) for the quantity mv^2 was coined by Leibniz (1646-1716). In the meantime the term *kinetic energy* has come into widespread use for one half of that quantity.

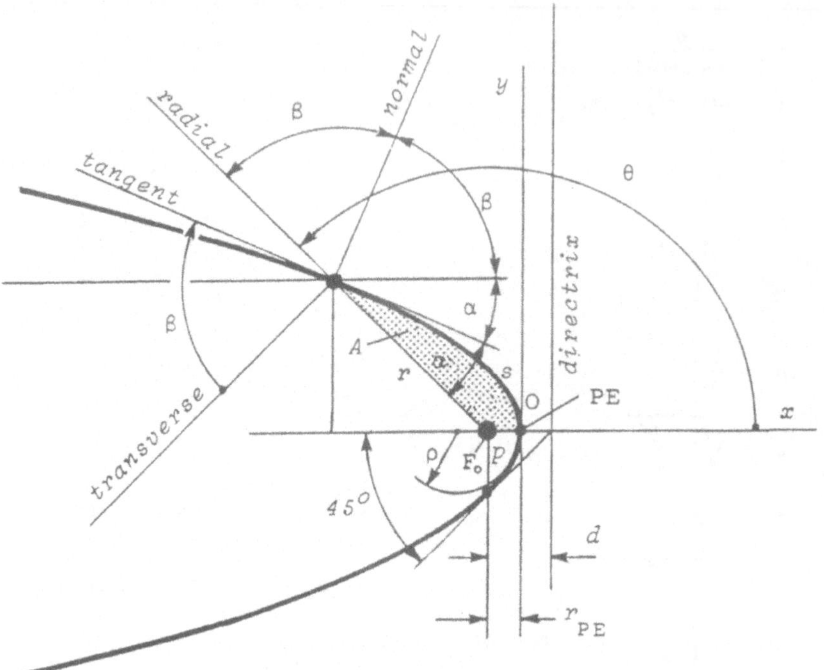

FIGURE 2.4. Parabola.

$$r = \frac{p}{1+\cos\theta} \qquad\qquad D = \sqrt{p}\,\tan\frac{\theta}{2}$$

$$y^2 = -2px \qquad\qquad \tan\beta = \frac{\sin\theta}{1+\cos\theta}$$

$$r_{PE} = \frac{p}{2} = q \qquad\qquad \rho = p$$

$$p = 2r_{PE} = d \qquad\qquad A = \frac{p^2}{6}\,\frac{2\sin\theta + \frac{1}{2}\sin 2\theta}{(1+\cos\theta)^2}$$

$$\varepsilon = 1 \qquad\qquad s = \frac{p}{2}\left[\frac{y}{p^2}\sqrt{p^2+y^2} + \ln\left[y+\sqrt{p^2+y^2}\right] - \ln p\right]$$

$$e = \infty \qquad\qquad a = \infty$$

TABLE 2.2. Key equations for parabolas.

For an elliptic orbit and a straight line segment path (free fall), it is

$$v^2 = \mu\left[\frac{2}{r} - \frac{1}{a}\right] \tag{2.10}$$

For a parabolic path it is

$$v^2 = \frac{2\mu}{r} \tag{2.11}$$

and for a hyperbolic path the vis-viva integral becomes

$$v^2 = \mu \left[\frac{2}{r} + \frac{1}{a'} \right] \tag{2.12}$$

The formulation

$$v = \sqrt{\mu \left[\frac{2}{r} - \frac{1}{a} \right]} \tag{2.13}$$

is often used, and referred to as the *speed equation*.

2.4   Velocities

There are a number of characteristic velocities which have special significance and special designations.

The *first cosmic velocity* (also called *circular velocity* or *Schuler velocity*) v_c is the velocity of a satellite on a circular orbit, whose radius is equal to the radius R of the surface of the master mass, which is assumed spherical. From the vis-viva integral (2.8) and with $\varepsilon - 0$, and $r = p = R$,

$$v_c = \sqrt{\frac{\mu}{R}} = \sqrt{gR} \tag{2.14}$$

For satellites of the Earth, using $\mu = 398\,601.19\ km^3/s^2$ and $R = 6371\ km$, the first cosmic velocity (which disregards the presence of the atmosphere) is

$$7.910\ km/s = 28\,475\ km/h$$

The *second cosmic velocity* (also called *parabolic velocity* or *escape velocity*) v_p is the velocity of a satellite on a parabolic path, as it passes the apsis, which is assumed located at a distance from the focus equal to the radius R of the master mass, which is again assumed to have a spherical shape.

From the vis-viva integral (2.11) for a satellite on a parabolic path

$$v_p = \sqrt{\frac{2\mu}{R}} = \sqrt{2gR} \tag{2.15}$$

The escape velocity for a satellite from the surface of the Earth (ignoring the presence of the atmosphere) is thus

$$11.186\ km/s = 40\,270\ km/h$$

For escape from the surface of the Moon the required escape velocity is

$$2.375\ km/s = 8\,548\ km/h$$

Satellites moving on a hyperbolic path still have a velocity when they reach infinity. This velocity is referred to as the *(hyperbolic) excess velocity* v_∞ and is obtained from the vis-viva integral (2.12) by setting $r = \infty$.

$$v_\infty = \sqrt{\frac{\mu}{a'}} \tag{2.16}$$

For a satellite on a parabolic path the excess velocity is

$$v_\infty = 0 \tag{2.17}$$

The velocity of a satellite on a given orbit reaches its lowest value when the satellite passes the apoapsis. The *apoapsis velocity*, for which $\cos \theta = -1$ (Figure 2.1), is given by

$$v_{AP} = \sqrt{\frac{\mu}{p}}\ (1 - \varepsilon) \tag{2.18}$$

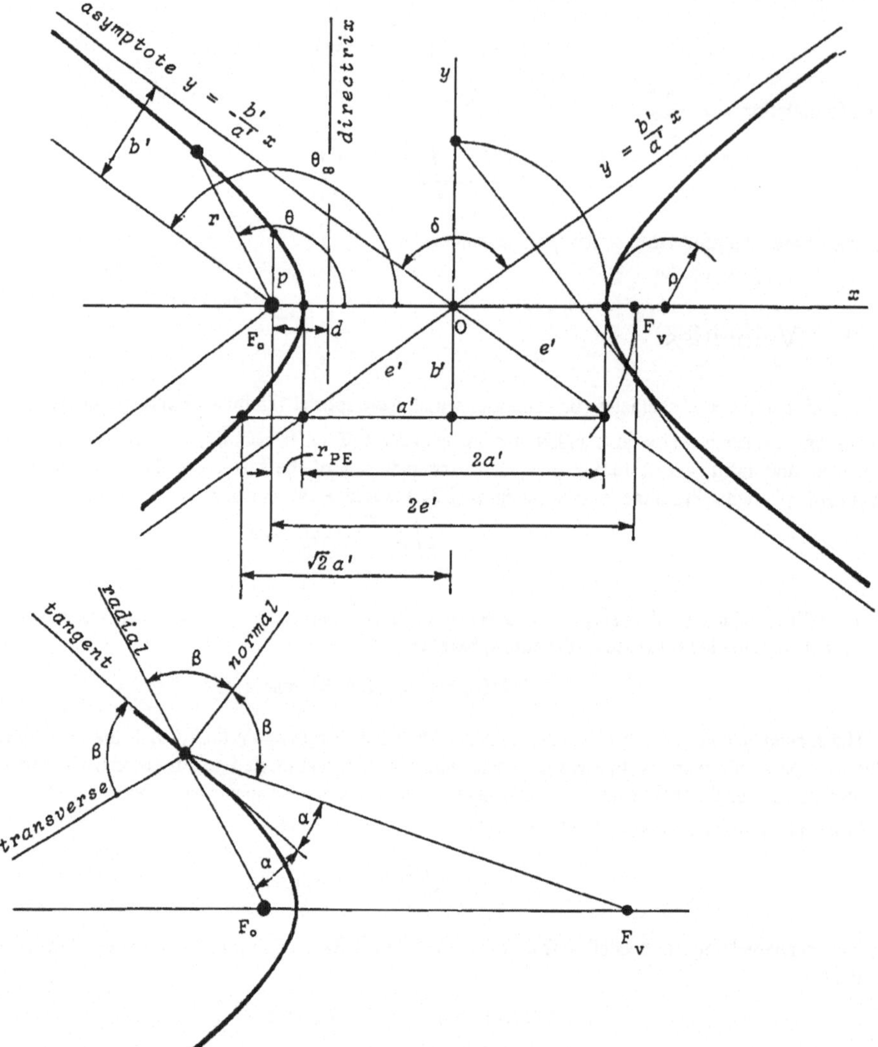

FIGURE 2.5. Hyperbola.

For a satellite on a straight line segment path ($p = 0, \varepsilon = 1$) we obtain

$$v_{AP} = 0 \tag{2.19}$$

For a satellite on an elliptic orbit, the apoapsis velocity can also be written

$$v_{AP} = \sqrt{\frac{\mu}{a}\frac{1-\varepsilon}{1+\varepsilon}} = \sqrt{\frac{\mu}{a}\frac{r_{PE}}{r_{AP}}} \tag{2.20}$$

For a parabolic path ($\varepsilon = 1$) it is

$$v_{AP} = 0 \tag{2.21}$$

The *periapsis velocity* is the highest velocity reached by a satellite on a given path. It is given by

$$v_{PE} = \sqrt{\frac{\mu}{p}}\,(1+\varepsilon) \tag{2.22}$$

$$r = \frac{p}{1 + \varepsilon \cos\theta}$$

$$a' = \frac{p}{\varepsilon^2 - 1}$$

$$b' = \frac{p}{\sqrt{\varepsilon^2 - 1}}$$

$$\frac{x^2}{a'^2} - \frac{y^2}{b'^2} = 1$$

$$p = \frac{b'^2}{a'} = \varepsilon d$$

$$\varepsilon = \sqrt{1 + \frac{b'^2}{a'^2}}$$

$$b' = \sqrt{a'p}$$

$$r_{PE} = \frac{p}{1 + \varepsilon} = a'(\varepsilon - 1) = q$$

$$p = a'(\varepsilon^2 - 1)$$

$$e' < a'$$

$$e' = \varepsilon a' = \frac{\varepsilon}{\varepsilon^2 - 1}p$$

$$e'^2 = a'^2 + b'^2$$

$$a' = -a$$

$$b' = jb$$

$$e' = -e$$

$$\tan\theta_\infty = \frac{b'}{a'} = -\sqrt{\varepsilon^2 - 1}$$

$$\cos\theta_\infty = -\frac{1}{\varepsilon}$$

$$1 < \varepsilon < \infty$$

$$a' = \frac{b'}{\sqrt{\varepsilon^2 - 1}}$$

$$\sin\frac{\delta}{2} = \frac{a'}{e'} = \frac{1}{\varepsilon}$$

$$\tan\beta = \frac{\varepsilon \sin\theta}{1 + \varepsilon \cos\theta}$$

$$\alpha + \beta = 90°$$

$$\rho = \frac{b'^2}{a'} = p$$

$$\mathrm{Sinh}\,E' = \frac{\sqrt{\varepsilon^2 - 1}\,\sin\theta}{1 + \varepsilon \cos\theta}$$

$$\mathrm{Cosh}\,E' = \frac{\varepsilon + \cos\theta}{1 + \varepsilon \cos\theta}$$

$$\sin\theta = \frac{\sqrt{\varepsilon^2 - 1}\,\mathrm{Sinh}\,E'}{\varepsilon\,\mathrm{Cosh}\,E' - 1}$$

$$\cos\theta = \frac{\varepsilon - \mathrm{Cosh}\,E'}{\varepsilon\,\mathrm{Cosh}\,E' - 1}$$

$$\tan\frac{\theta}{2} = \sqrt{\frac{\varepsilon^2 + 1}{\varepsilon^2 - 1}}\,\mathrm{Tanh}\frac{E'}{2}$$

$$r = a'(\varepsilon\,\mathrm{Cosh}\,E' - 1)$$

$$M' = \varepsilon\,\mathrm{Sinh}\,E' - E'$$

TABLE 2.3. Key equations for hyperbolas.

A satellite on a straight line segment path ($p = 0$, $\varepsilon = 1$) represents a limiting case. Its periapsis velocity would occur at the moment where the satellite passes through the point-size master mass (i.e. the periapsis of its path coincides with the mass centre of the master mass).

$$v_{PE} = \infty$$

For a satellite on an elliptic path the periapsis velocity can be expressed by

$$v_{PE} = \sqrt{\frac{\mu}{a}\frac{1 + \varepsilon}{1 - \varepsilon}} = \sqrt{\frac{\mu}{a}\frac{r_{AP}}{r_{PE}}} \tag{2.23}$$

on a parabolic path

$$v_{PE} = 2\sqrt{\frac{\mu}{p}} \tag{2.24}$$

and on a hyperbolic path

$$v_{PE} = \sqrt{\frac{\mu}{a'}\frac{\varepsilon + 1}{\varepsilon - 1}} \tag{2.25}$$

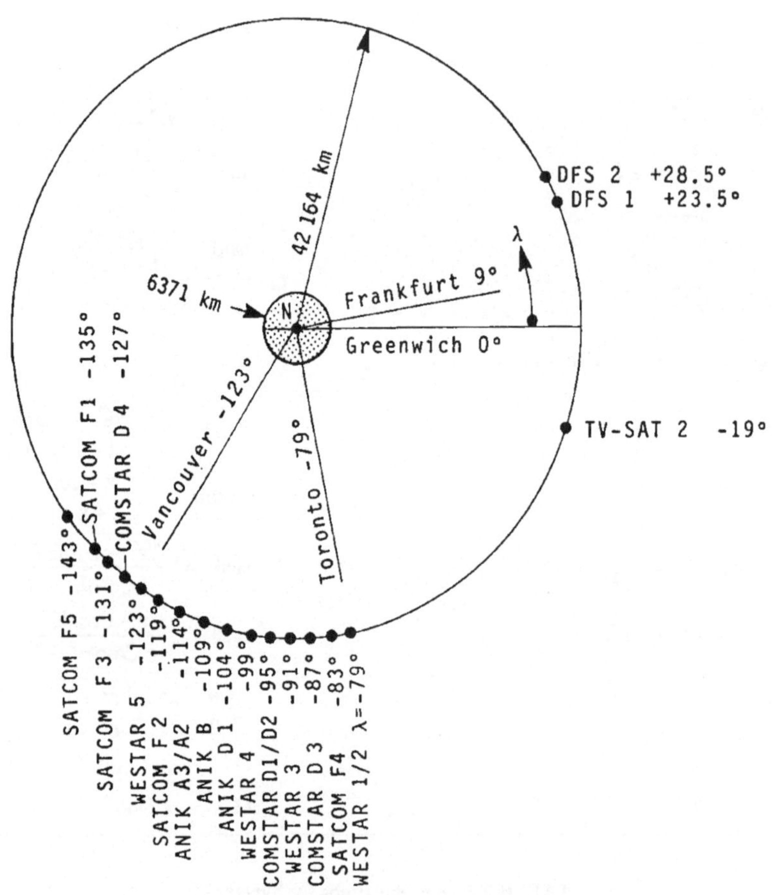

FIGURE 2.6. Some of the geosynchronous satellite locations.

The *mean angular velocity n*, equation (2.4), and the *mean anomaly* M, equation (2.44), are related by

$$n = \frac{dM}{dt} \tag{2.26}$$

and

$$M = n(t - t_{PE}) \tag{2.27}$$

where t is the time, and t_{PE} the time at periapsis passage. Equation (2.27) is known as the *Kepler equation*.

2.5   Potential Energy

The work done by the Kepler force $\mathbf{K}$ on a point satellite m as it is displaced from a position $r = r$ to a position $r = \infty$ (Figure 2.7) is given by

$$\int\limits_{r}^{\infty} \mathbf{K} \cdot d\mathbf{r} = -\int\limits_{r}^{\infty} \frac{\mu m}{r^2}\, dr = \left.\frac{\mu m}{r}\right|_{r}^{\infty} = -\frac{\mu m}{r} \tag{2.28}$$

The potential energy

$$U = -\frac{\mu m}{r} \tag{2.29}$$

of a point satellite in the gravitational field of a point master is equal to the work done by the Kepler force as it is displaced from a given position r to infinity.

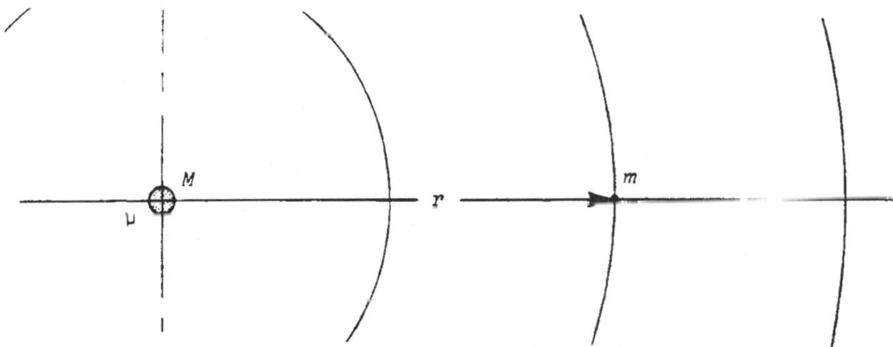

FIGURE 2.7. A point satellite in the gravitational field of its master.

   The potential energy is thus always negative. It vanishes at infinity. It reaches a value of negative infinity at $r = 0$, i.e. at the mass centre of the master mass. For astrodynamics, only the region outside the master mass is of practical interest. At the surface of a master mass of spherical shape (for which the potential energy equation outside the master mass is the same as that for a point-sized master mass, as will be shown later) the potential energy assumes the value

$$-\frac{\mu m}{R}$$

where R is the radius of the spherical master mass. If one takes into consideration the fact that the shape of the Earth and other celestial bodies deviates somewhat from a perfect sphere, correction terms will appear in the expression for the potential energy. They will be discussed later.

   For the Earth sphere, with $R = 6371$ km and $\mu = 398\,601.19$ km^3/s^2, the potential energy of a satellite of unit mass located immediately above the Earth surface is

$$-\frac{\mu}{R} = -62.565\, \frac{\text{km}^2}{\text{s}^2}$$

For the Moon (which is essentially spherical), with $R = 1739$ km and $\mu = 4902.73$ km^3/s^2, the potential energy of a satellite of unit mass located immediately above the Moon surface amounts to

$$-\frac{\mu}{R} = -2.819\, \frac{\text{km}^2}{\text{s}^2}$$

Figure 2.8 shows the potential energy as function of distance from Earth centre.

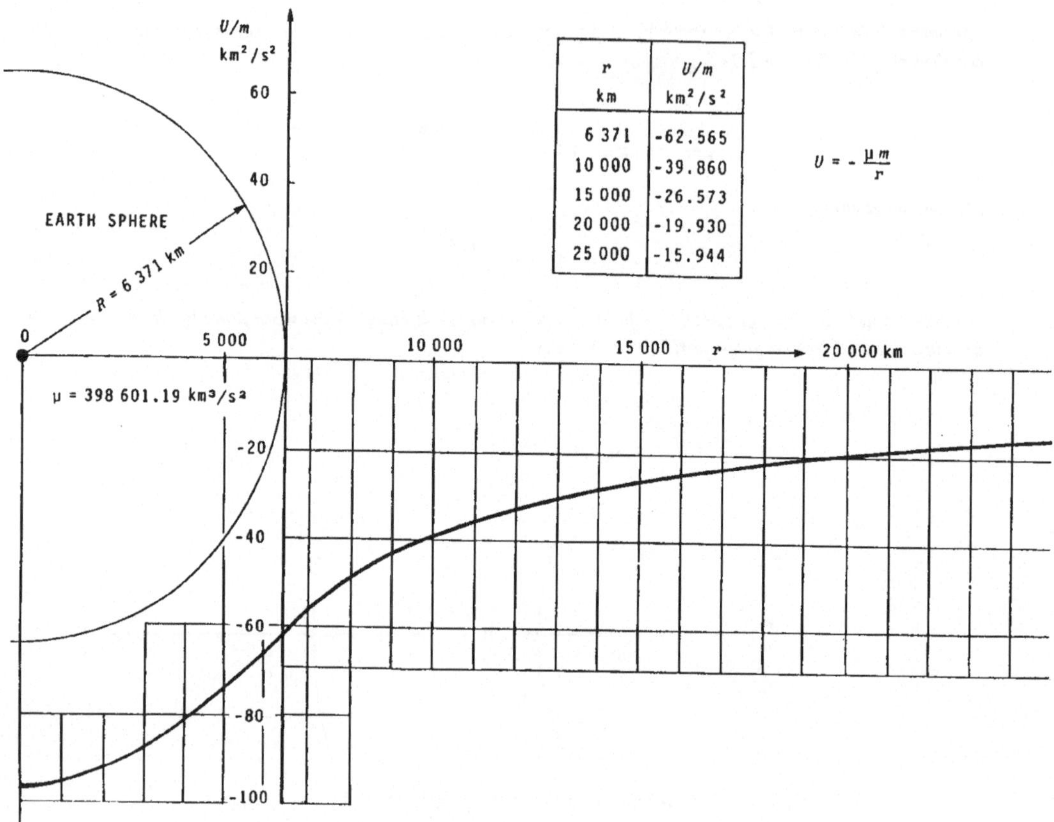

r	U/m
km	km²/s²
6 371	-62.565
10 000	-39.860
15 000	-26.573
20 000	-19.930
25 000	-15.944

$$U = -\frac{\mu m}{r}$$

FIGURE 2.8. Potential energy of a satellite of the Earth sphere.

The attraction force $\mathbf{K}$ acting on satellite m is the negative gradient of the potential energy

$$\mathbf{K} = -\mathbf{grad}\, U \qquad (2.30)$$

Inside a homogeneous sphere, the potential energy can be shown to be

$$U = -\frac{\mu m}{2R}\left[3 - \frac{r^2}{R^2}\right] \qquad (2.31)$$

and the attraction force acting on a point mass m inside such sphere is

$$\mathbf{F} = -\mathbf{grad}\, U = -\frac{\mu m}{R^3} r\, \mathbf{e}_r$$

For a plot of the potential energy inside the Earth see also Figure 2.8.

Inside a *hollow* sphere, the potential energy is zero throughout.

$$U = 0 \qquad (2.32)$$

2.6 Total Energy

The *potential energy* of a point satellite m in the gravitational field of a point (or spherical) master mass is, according to equation (2.29),

$$U = -\frac{\mu m}{r}$$

The *kinetic energy* of a point satellite of mass m is given by the well-known formula

$$T = \frac{mv^2}{2}$$

Using the vis-viva integral (2.8), the kinetic energy can also be expressed by

$$T = \frac{\mu m}{2}\left[\frac{2}{r} - \frac{1-\varepsilon^2}{p}\right] \tag{2.33}$$

The total energy is the sum of kinetic and potential energy.

$$E = T + U = \frac{\mu m}{2}\left[\frac{2}{r} - \frac{1-\varepsilon^2}{p}\right] - \frac{\mu m}{r}$$

or

$$E = -\frac{\mu m(1-\varepsilon^2)}{2p} \tag{2.34}$$

If the shape of the satellite's orbit is specified, the expressions for the total energy can be written very compactly.

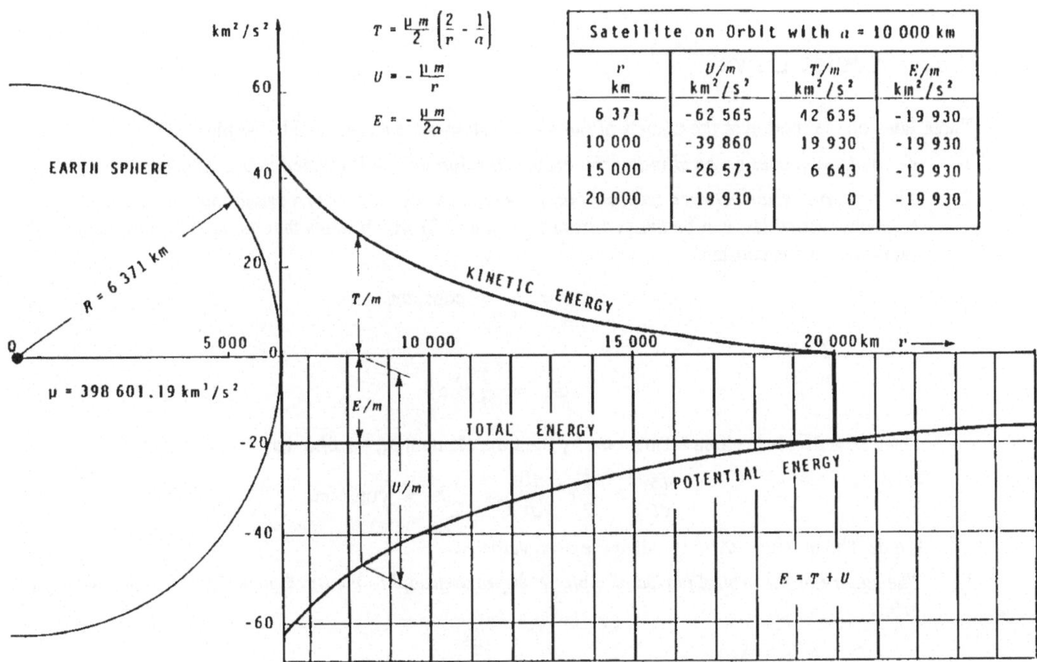

Satellite on Orbit with a = 10 000 km			
r km	U/m km²/s²	T/m km²/s²	E/m km²/s²
6 371	-62 565	42 635	-19 930
10 000	-39 860	19 930	-19 930
15 000	-26 573	6 643	-19 930
20 000	-19 930	0	-19 930

FIGURE 2.9. Total energy of a point satellite of the Earth.

The total energy of a point satellite on an elliptic or circular or straight-line orbit is always negative (Figure 2.9).

$$E \; = \; -\frac{\mu m}{2a} \tag{2.35}$$

On a parabolic orbit, a satellite's total energy vanishes.

$$E = 0 \tag{2.36}$$

On a hyperbolic orbit, the total energy is always positive.

$$E \; = \; \frac{\mu m}{2a'} \tag{2.37}$$

The *specific mechanical energy* e is the total energy per unit mass:

$$e \; = \; \frac{E}{m}$$

A spacecraft resting on the Earth sphere's surface (neglecting the velocity due to the Earth's spin) has a specific mechanical energy that consists of potential energy only and amounts to

$$e \; = \; -\frac{\mu}{R} \; = \; -\frac{398\,601}{6371} \; = \; -62.565 \; \text{MJ/kg}$$

In a Low Earth Orbit (LEO) of, say, $r \; = \; 6571$ km radius, a point satellite's specific mechanical energy is

$$e \; = \; -\frac{\mu}{2r} \; = \; -\frac{398\,601}{2(6571)} \; = \; -30.330 \; \text{MJ/kg}$$

and in a Geosynchronous Orbit (GEO), with $r \; = \; 42\,164$ km, a point satellite's specific mechanical energy is

$$e \; = \; -\frac{398\,601}{2(42\,164)} \; = \; -4.726 \; \text{MJ/kg}$$

2.7   Kepler Laws

Three laws on the motion of the planets of our solar system were enunciated by Kepler:

1. *Each planet moves on an elliptic orbit around the Sun, which is located in one of the foci.*

2. *The position vector between Sun and planet sweeps out equal areas in equal time intervals.*
   Kepler's second law can be derived from equation (2.2) which states that the angular momentum of a satellite remains constant.

$$r^2\dot\theta \; = \; h \; = \; \text{constant}$$

   From Figure 2.1

$$dA \; = \; \frac{1}{2}r^2 \, d\theta$$

   If we now divide the area element dA by the time element dt, we find that

$$\frac{dA}{dt} \; = \; \frac{1}{2}r^2 \frac{d\theta}{dt} \; = \; \frac{1}{2}h \; = \; \text{constant}$$

   q.e.d. The quantity dA/dt is called the *area velocity*.

3. *The square of the orbital period of a planet is proportional to the third power of the major axis of its orbit.*

$$\tau^2 \; = \; \frac{4\pi^2}{\mu} \, a^3$$

   as can be seen from equation (2.3).

2.8 Time of Flight

Neither the radius r nor the polar angle θ of the position of a satellite on a Kepler orbit can be expressed explicitly and in closed form as a function of time, i.e. as

$$r = r(t)$$

or

$$\theta = \theta(t)$$

On the other hand, it is possible to express the time explicitly and in closed form either as function of the polar angle θ or also as function of the radius vector; e.g. as

$$t = t(\theta)$$

We proceed by using the specific angular momentum, equation (2.2),

$$r^2 \dot{\theta} = \sqrt{\mu p}$$

and replace the radius r by means of equation (2.1)

$$\frac{d\theta}{dt} = \sqrt{\frac{\mu}{p^3}} \, (1 + \varepsilon \cos\theta)^2$$

Separation of the variables

$$dt = \sqrt{\frac{p^3}{\mu}} \, \frac{d\theta}{(1 + \varepsilon \cos\theta)^2} \tag{2.38}$$

and integration between the limits t_{PE} (= time of periapsis passage) and t on the left hand side and 0 and θ on the right hand side results in

$$t - t_{PE} = \sqrt{\frac{p^3}{\mu(1 - \varepsilon^2)^3}} \left[\arcsin \frac{\sqrt{1 - \varepsilon^2} \, \sin\theta}{1 + \varepsilon \cos\theta} - \varepsilon \frac{\sqrt{1 - \varepsilon^2} \, \sin\theta}{1 + \varepsilon \cos\theta} \right] \tag{2.39}$$

or, since $j \sin\theta = \mathrm{Sinh} j\theta$

$$t - t_{PE} = \sqrt{\frac{p^3}{\mu(\varepsilon^2 - 1)^3}} \left[\varepsilon \frac{\sqrt{\varepsilon^2 - 1} \, \sin\theta}{1 + \varepsilon \cos\theta} - \mathrm{ArSinh} \frac{\sqrt{\varepsilon^2 - 1} \, \sin\theta}{1 + \varepsilon \cos\theta} \right] \tag{2.40}$$

Equation (2.39) expresses the time of flight in a form suitable for elliptic orbits, while equation (2.40) is suitable for hyperbolic orbits.

For parabolic orbits, i.e. for the special case $\varepsilon = 1$, equation (2.38) leads to

$$t - t_{PE} = \sqrt{\frac{p^3}{\mu}} \int_0^\theta \frac{d\theta}{(1 + \cos\theta)^2}$$

which, upon integration, becomes

$$t - t_{PE} = \frac{1}{3} \sqrt{\frac{p^3}{\mu}} \, \frac{(2 + \cos\theta)\sin\theta}{(1 + \cos\theta)^2} \tag{2.41}$$

2.9   The Anomalies

In space mechanics there are three different *anomalies* in use.

The *true anomaly* θ is the angle between the line of apsides of the satellite orbit and the position vector of the satellite, as shown in Figure 2.1.

The *eccentric anomaly* E is related to the true anomaly by

$$\sin E \;=\; \frac{\sqrt{1 - \varepsilon^2}\,\sin\theta}{1 + \varepsilon\,\cos\theta} \tag{2.42}$$

or by

$$\tan\frac{E}{2} \;=\; \sqrt{\frac{1 - \varepsilon}{1 + \varepsilon}}\;\tan\frac{\theta}{2} \tag{2.43}$$

The eccentric anomaly is the angle swept out at the centre point C by a point Q moving along the circumscribed auxiliary circle such that the point P on the ellipse and point Q lie on a straight line perpendicular to the line of apsides (Figure 2.10).

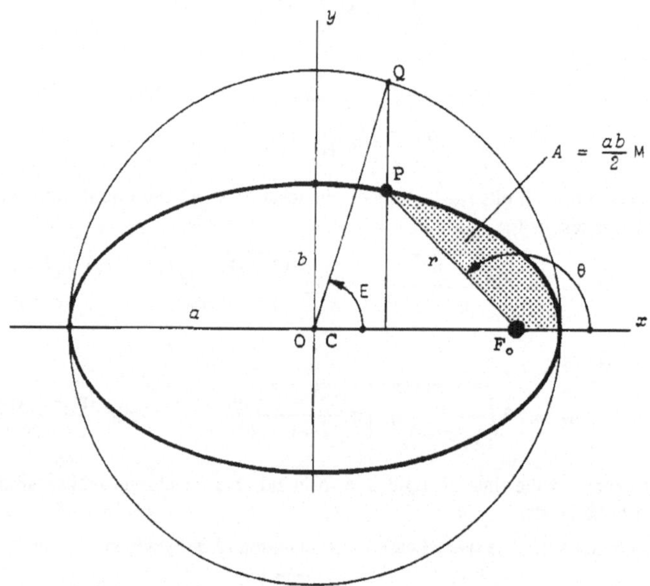

FIGURE 2.10. True, eccentric, and mean anomalies.

The combination ($E - \varepsilon \sin E$) occurs in the Kepler equation (2.27). For it we introduce the term *mean anomaly* and the symbol M

$$M \;=\; E - \varepsilon\,\sin E \tag{2.44}$$

The mean anomaly is the angle swept out by the radius vector of an imagined satellite moving at the imagined *constant* angular speed $n = 2\pi/\tau$ (instead of the satellite's actual angular speed $\dot\theta = d\theta/dt$).

$$M \;=\; \frac{2\pi}{\tau}(t - t_{PE}) \;=\; n(t - t_{PE}) \tag{2.45}$$

For parabolic orbits, the terms eccentric anomaly and mean anomaly become meaningless.

For hyperbolic orbits, quantities corresponding to the eccentric anomaly and the mean anomaly can be defined. There are sometimes called *hyperbolic eccentric anomaly* and *hyperbolic mean anomaly*, and are defined by

$$\text{Sinh}E' = \frac{\sqrt{\varepsilon^2 - 1}\ \sin\theta}{1 + \varepsilon\cos\theta} \tag{2.46}$$

or

$$\text{Tanh}\frac{E'}{2} = \sqrt{\frac{\varepsilon - 1}{\varepsilon + 1}}\ \tan\frac{\theta}{2} \tag{2.47}$$

and by

$$M' = \varepsilon\,\text{Sinh}E' - E' \tag{2.48}$$

See also Figure 2.11.

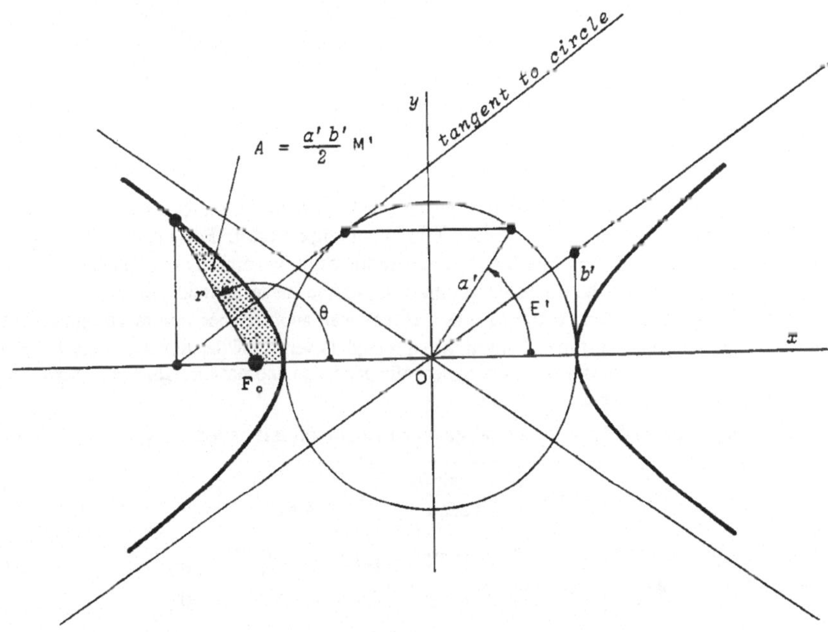

FIGURE 2.11. The hyperbolic anomalies.

It can readily be shown that

$$E' = -jE$$

and that

$$M' = jM$$

2.10 A Universal Variable

The use of the anomalies becomes awkward when near-parabolic orbits are to be studied. To avoid these difficulties, another variable is then used. It is called the *universal variable* and commonly expressed by the symbol x.

The universal variable is defined by

$$x = \sqrt{a}\,(E - E_o) \tag{2.49}$$

where E represents the eccentric anomaly. For parabolic orbits

$$x = \sqrt{p}\left[\tan\frac{\theta}{2} - \tan\frac{\theta_o}{2}\right] \tag{2.50}$$

and for hyperbolic orbits

$$x = \sqrt{a'}\,(E' - E'_o) \tag{2.51}$$

where E_o, E'_o, and θ_o refer to initial values occurring at time $t = 0$. If $t = 0$ coincides with the satellite's passage through the periapsis of its orbit, then $E_o = E'_o = \theta_o = 0$.

For a fuller treatment of the implementation and use of the universal variable formulation see e.g. Bate et al. (1971).

2.11 Kepler Equations

In order to express the time of flight of a satellite, measured from the time $t_{\rm PE}$ of its passage through the periapsis, we make use of equation (2.39) for a satellite on an elliptic orbit. For a satellite on a parabolic orbit, we use equation (2.41), and for a satellite on a hyperbolic orbit, we employ equation (2.40). We can use either the true anomaly θ, or the eccentric anomaly E, or the mean anomaly M as the independent variable. For a parabolic path, eccentric anomaly as well as mean anomaly become meaningless (they are both zero throughout) and thus the time of flight is expressed in terms of the true anomaly only. For a hyperbolic path either the true anomaly θ, or the hyperbolic eccentric anomaly E', or the hyperbolic mean anomaly M' can be used as independent variable.

The equation giving the time of flight for a satellite on an elliptic orbit is called the *Kepler equation*.

$$t - t_{\rm PE} = \sqrt{\frac{a^3}{\mu}}\left[\arcsin\frac{\sqrt{1-\varepsilon^2}\,\sin\theta}{1+\varepsilon\cos\theta} - \varepsilon\,\frac{\sqrt{1-\varepsilon^2}\,\sin\theta}{1+\varepsilon\cos\theta}\right] \tag{2.52a}$$

$$t - t_{\rm PE} = \sqrt{\frac{a^3}{\mu}}\left[2\arctan\left[\sqrt{\frac{1-\varepsilon}{1+\varepsilon}}\,\tan\frac{\theta}{2}\right] - \varepsilon\,\frac{\sqrt{1-\varepsilon^2}\,\sin\theta}{1+\varepsilon\cos\theta}\right] \tag{2.52b}$$

$$t - t_{\rm PE} = \sqrt{\frac{a^3}{\mu}}\,[E - \varepsilon\sin E] \tag{2.52c}$$

$$t - t_{\rm PE} = \sqrt{\frac{a^3}{\mu}}\,M \tag{2.52d}$$

Kepler derived the equation now bearing his name when he was studying the motion of planets. Thus he was concerned with elliptical orbits only. The same type of equation is, however, also applicable to parabolic and hyperbolic orbits.

For parabolic orbits the time of flight equation is also referred to as the *Olbers equation*[*].

$$t - t_{\rm PE} = \frac{1}{3}\sqrt{\frac{p^3}{\mu}}\,\frac{(2+\cos\theta)\sin\theta}{(1+\cos\theta)^2} \tag{2.53a}$$

$$t - t_{\rm PE} = \frac{1}{2}\sqrt{\frac{p^3}{\mu}}\left[\tan\frac{\theta}{2} + \frac{1}{3}\tan^3\frac{\theta}{2}\right] \tag{2.53b}$$

---

[*]Wilhelm Olbers (1758-1840), German astronomer.

For a hyperbolic orbit the time of flight is given by

$$t - t_{PE} = \sqrt{\frac{a'^3}{\mu}} \left[\varepsilon \frac{\sqrt{\varepsilon^2 - 1} \sin\theta}{1 + \varepsilon \cos\theta} - \text{ArSinh} \frac{\sqrt{\varepsilon^2 - 1} \sin\theta}{1 + \varepsilon \cos\theta} \right] \tag{2.54a}$$

$$t - t_{PE} = \sqrt{\frac{a'^3}{\mu}} \left[\varepsilon \frac{\sqrt{\varepsilon^2 - 1} \sin\theta}{1 + \varepsilon \cos\theta} - \ln \frac{\sqrt{\varepsilon + 1} + \sqrt{\varepsilon - 1} \tan\frac{\theta}{2}}{\sqrt{\varepsilon + 1} - \sqrt{\varepsilon - 1} \tan\frac{\theta}{2}} \right] \tag{2.54b}$$

$$t - t_{PE} = \sqrt{\frac{a'^3}{\mu}} \left[\varepsilon \, \text{Sinh} \, E' - E' \right] \tag{2.54c}$$

$$t - t_{PE} = \sqrt{\frac{a'^3}{\mu}} \, M' \tag{2.54d}$$

If one replaces the true anomaly θ by the radius r in the preceding equations, then the time of flight can be expressed in the form

$$t = t(r)$$

This formulation also permits the time of flight prediction of a satellite on a straight line segment path.

With the help of equation (2.1) we can write

$$\frac{\sqrt{1 - \varepsilon^2} \sin\theta}{1 + \varepsilon \cos\theta} \div \frac{r}{p} \frac{\sqrt{1 - \varepsilon^2}}{\varepsilon} \sqrt{\varepsilon^2 - \left[\frac{p}{r} - 1 \right]^2}$$

and obtain for an elliptic or straight line segment (= free fall) orbit

$$t - t_{PE} = \sqrt{\frac{a^3}{\mu}} \left[\arcsin \frac{\sqrt{2ar - a^2(1 - \varepsilon^2) - r^2}}{\varepsilon a} - \frac{1}{a} \sqrt{2ar - a^2(1 - \varepsilon^2) - r^2} \right] \tag{2.55a}$$

For a parabolic orbit

$$t - t_{PE} = \frac{1}{3} \sqrt{\frac{p^3}{\mu}} \left[\frac{r}{p} + 1 \right] \sqrt{2 \frac{r}{p} - 1} \tag{2.55b}$$

and for a hyperbolic orbit

$$t - t_{PE} = \sqrt{\frac{a'^3}{\mu}} \left[\frac{1}{a'} \sqrt{2a'r - a'^2(\varepsilon^2 - 1) + r^2} - \text{ArSinh} \frac{\sqrt{2a'r - a'^2(\varepsilon^2 - 1) + r^2}}{\varepsilon a'} \right] \tag{2.55c}$$

2.12 Lambert Equations

For a point satellite travelling on an elliptic orbit from a position 1 to a position 2, one can determine the time interval $t_2 - t_1$ for the duration of the flight from position 1 to position 2 by means of the Kepler equation (2.52c).

$$t_2 - t_1 = \sqrt{\frac{a^3}{\mu}} \left[E_2 - \varepsilon \sin E_2 - (E_1 - \varepsilon \sin E_1) \right] \tag{2.56}$$

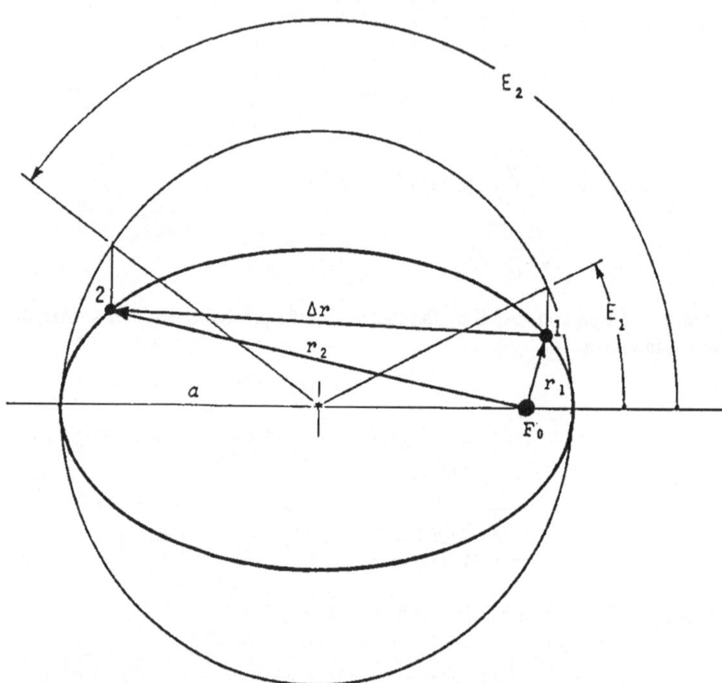

FIGURE 2.12. Lambert's equation.

Employing a trigonometric identity, equation (2.56) may also be written

$$t_2 - t_1 = \sqrt{\frac{a^3}{\mu}} \left[E_2 - E_1 - 2\varepsilon \sin \frac{E_2 - E_1}{2} \cos \frac{E_2 + E_1}{2} \right] \qquad (2.57)$$

For the case where the eccentric anomalies E_1 and E_2 are unknown, but the radii r_1 and r_2 and the distance Δr between position 1 and 2 are known, Lambert[*] has shown that the use of two auxiliary parameters α and β permits a direct determination of the time interval. He introduced (Figure 2.12)

$$E_2 - E_1 = \alpha - \beta \qquad (2.58)$$

and

$$\varepsilon \cos \frac{E_2 + E_1}{2} = \cos \frac{\alpha + \beta}{2} \qquad (2.59)$$

and obtained an expression which is now known as the *Lambert equation* for an elliptic orbit

$$t_2 - t_1 = \sqrt{\frac{a^3}{\mu}} \left[\alpha - \sin\alpha - (\beta - \sin\beta) \right] \qquad (2.60)$$

The auxiliary parameters α and β are defined by

$$\sin \frac{\alpha}{2} = \frac{1}{2} \sqrt{\frac{r_1 + r_2 + \Delta r}{a}} \qquad (2.61)$$

---

[*]Johann Heinrich Lambert (1728-1777), German physicist and philosopher.

$$\sin \frac{\beta}{2} = \frac{1}{2} \sqrt{\frac{r_1 + r_2 - \Delta r}{a}} \tag{2.62}$$

It can be shown that for a parabolic orbit, Lambert's equation yields

$$t_2 - t_1 = \frac{1}{6\sqrt{\mu}} \left[\sqrt{(r_1 + r_2 + \Delta r)^3} - \sqrt{(r_1 + r_2 - \Delta r)^3} \right] \tag{2.63}$$

For a hyperbolic orbit, Lambert's equation assumes the form

$$t_2 - t_1 = \sqrt{\frac{a'^3}{\mu}} \left[\text{Sinh}\alpha' - \alpha' - (\text{Sinh}\beta' - \beta') \right] \tag{2.64}$$

where the auxiliary parameters α' and β' are defined by

$$\text{Sinh} \frac{\alpha'}{2} = \frac{1}{2} \sqrt{\frac{r_1 + r_2 + \Delta r}{a'}} \tag{2.65}$$

$$\text{Sinh} \frac{\beta'}{2} = \frac{1}{2} \sqrt{\frac{r_1 + r_2 - \Delta r}{a'}} \tag{2.66}$$

Lambert's equations do not indicate the direction of flight. Obviously the flight interval is always positive. Positions 1 and 2 must thus be selected such that the satellite proceeds from position 1 to position 2, regardless of whether it moves towards or away from the periapsis of its orbit.

2.13 Orbital Elements

The use of an absolute (i.e. space-fixed) Cartesian coordinate system, at whose origin the master mass is located, results in a system of three coupled differential equations for equation (1.20).

$$\ddot{x} = -\frac{\mu x}{(x^2 + y^2 + z^2)^{3/2}} \tag{2.67a}$$

$$\ddot{y} = -\frac{\mu y}{(x^2 + y^2 + z^2)^{3/2}} \tag{2.67b}$$

$$\ddot{z} = -\frac{\mu z}{(x^2 + y^2 + z^2)^{3/2}} \tag{2.67c}$$

An inspection of this set of equations indicates that integration will require *six* integration constants. They are known as *orbital elements* and are needed to specify a satellite orbit, its location and attitude with respect to the master mass, including the civic time of the satellite's passage through a characteristic point on its orbit.

Basically it does not matter which six constants are chosen as fundamental, because other orbital constants can be obtained readily in terms of the fundamental ones.

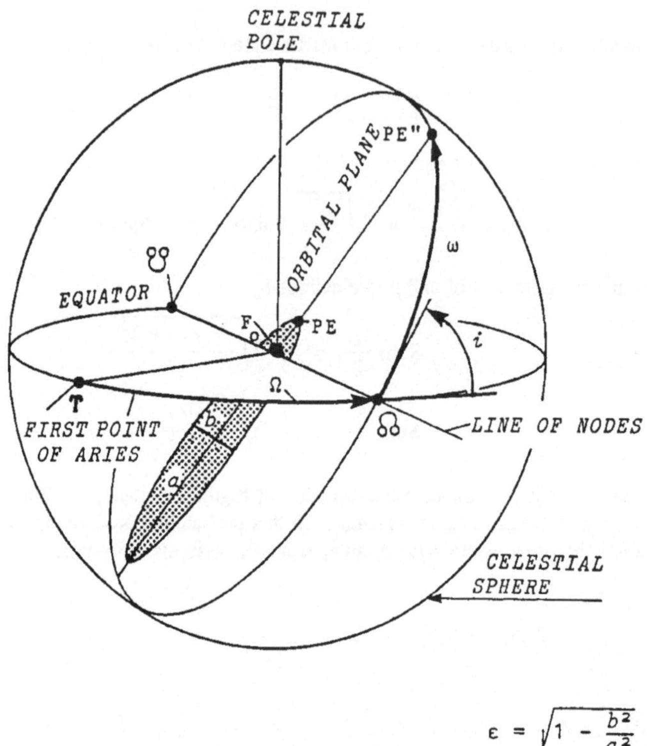

$$\varepsilon = \sqrt{1 - \frac{b^2}{a^2}}$$

FIGURE 2.13. The geometrical orbital elements.

It is widespread practice to use the following as *fundamental* orbital elements (or *Keplerian elements*)

$$a,\ \varepsilon,\ \Omega,\ \omega,\ i,\ t_{\mathrm{PE}}$$

Thus there are five geometrical elements and one time element. The five geometrical elements (Figure 2.13) may again be subdivided into two shape elements and three position elements. We then have

the shape elements

$a =$ semi-major axis

$\varepsilon =$ numerical eccentricity

the position elements

$\Omega =$ angle of right ascension of ascending node

$i =$ inclination angle of orbital plane

$\omega =$ angle from ascending node to periapsis PE, (argument of periapsis)

and the time element

$t_{\mathrm{PE}} =$ point in time ("epoch") of satellite's passage through periapsis

The reader's attention is drawn to the unfortunate circumstance that in astronomy, the use of the symbols ω and Ω for the two specific *angles* mentioned above is so entrenched that these symbols have become irreplaceable. Just as irreplaceable are the same symbols in mechanics, where they represent *angular velocities*.

2.14 Tethered Satellites

Engineers speak of *tethered satellites* (Misra and Modi, 1983) when two satellites are connected by a cable. In the subsequent treatment of the problem, it is assumed that each of the satellites is of *point* size, and that the connecting cable is *massless*. In Figure 2.14 a pair of tethered point satellites is shown, moving on circular orbits in the gravitational field of a point master M. For satellite m_1 the Kepler force is reduced by the cable tension T, such that

$$K_1 \; - \; T \; = \; \frac{\mu m_1}{r_1^2} \; - \; T$$

is pointing towards the master. An adaptation of equation (2.9) leads to the satellite's velocity

$$v_1 \; = \; \sqrt{\frac{\mu}{r_1} \; - \; \frac{T r_1}{\mu_1}} \tag{2.68}$$

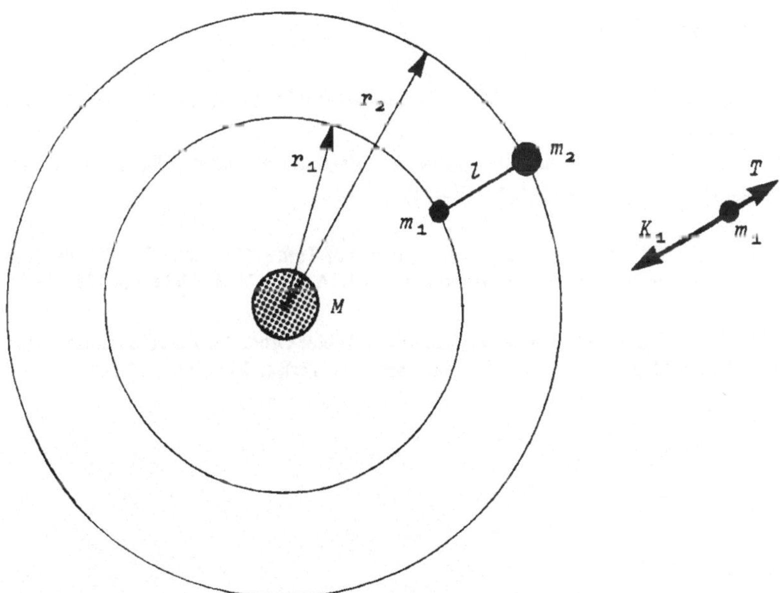

FIGURE 2.14. Tethered satellites.

Satellite m_2 is subjected to a force

$$K_2 \; + \; T \; = \; \frac{\mu m_2}{r_2^2} \; + \; T$$

pointing towards the master. The satellite moves at a velocity of

$$v_2 \; = \; \sqrt{\frac{\mu}{r^2} \; + \; \frac{T r_2}{m^2}} \tag{2.79}$$

From the condition, that the orbital angular velocity Ω of the two satellites is the same,

$$\Omega \; = \; \frac{v_1}{r_1} \; = \; \frac{v_2}{r_2} \tag{2.80}$$

$$\Omega \;=\; \frac{v_1}{r_1} \;=\; \frac{v_2}{r_2} \tag{2.80}$$

the cable tension T can be determined. It amounts to

$$T \;=\; \frac{\mu m_1 m_2 (r_2^3 - r_1^3)}{r_1^2 r_2^2 (m_1 r_1 + m_2 r_2)} \tag{2.81}$$

When both satellites have the same mass, i.e. when $m_1 = m_2 = m$, and when the cable length l is very short, i.e. when $l/r_1 \ll 1$, and $r_1 = r$ and $r_2 = r + l$, then the cable tension becomes simply

$$T \;=\; \frac{3}{2} \frac{\mu m l}{r^3} \tag{2.82}$$

As an example, consider two tethered satellites of 100 kg mass each, connected by a massless cable of 1000 m length, on a circular orbit of 7000 km radius about the Earth. The cable tension is

$$T \;=\; \frac{3}{2} \frac{398\,601\,(100)\,1000}{7000^3} \;=\; 0.174\,\text{N}$$

Suggested Reading

1. Bate, R.R.; D.D. Mueller; J.C. White. *Fundamentals of Astrodynamics*, Dover, 1971, 455 pp.

2. Magnus, K. "Kreisel als vielseitige Mittel in Luft- und Raumfahrt", *Zeitschrift für Flugwissenschaften und Weltraumforschung*, 2, 4, 1978, pp. 271-227.

3. Misra, A.K.; V.J. Modi. "Dynamics and Control of Tether-Connected Two-Body Systems - A Brief Review", *Space 2000* (Editor: L. Napolitano), AIAA, New York, 1983, pp. 473-514.

4. Schiehlen, W.; O. Kolbe. "A Generalization of Kepler Orbits for Large Satellites", *Applied Mechanics* (Editors: M. Hetényi; W.G. Vincenti), Springer-Verlag, 1969, pp. 364-381.

Problems

2.1   Determine the orbital periods for the following three satellites, which all orbit the Earth. (a) $a = 7000$ km, $\varepsilon = 0.1$; (b) $b = 7000$ km, $\varepsilon = 0.1$; and (c) $p = 7000$ km, $\varepsilon = 0.1$. Sketch the orbits and discuss the results.

2.2   Determine the Schuler period for Sun, Earth, Moon, Mars and Jupiter.

2.3   What is (a) the first cosmic, and (b) the second cosmic speed of a satellite of the Sun, Earth, Moon, Mars and Jupiter?

2.4   A satellite is orbiting the Earth. Its orbital ellipse has a semi-major axis of 8000 km. What is the satellite's speed when it is at a distance of 8500 km from Earth centre? Make a sketch, with $r_{\text{PE}} = 7000$ km.

2.5 A satellite moves on an orbit with $a = 10\,000$ km and $\varepsilon = 0.6$. (a) Make a sketch; (b) Determine the eccentric and mean anomalies for the true anomalies given, and enter them into the table shown below; and (c) What is the significance of $\theta = 126.87°$, and $\theta = 147.69°$?

True Anomaly θ		Eccentric Anomaly E		Mean Anomaly M	
°	rad	°	rad	°	rad
0	0	0	0	0	0
45					
90	π/2				
126.87					
135					
147.69					
150					
180	π		π		π
270					
360	2π	360	2π	360	2π

2.6 The Schuler period (of 84 min for the Earth) provides a basis for the removal of so-called ballistic disturbances (due to vehicle manoeuvring) from gyroscopic navigation instruments. In order to get an idea of the significance of the Schuler period, calculate (a) the period of a mathematical pendulum suspended from the Earth surface with a length equal to the Earth radius; (b) the period of motion of a satellite moving in a shaft drilled through the Earth centre, or (c) in an arbitrary fashion; (d) the period of a point moving on a rail tangent to the Earth surface; (e) a satellite skimming along the surface of the Earth sphere. (The gravitational force inside a homogeneous sphere is $\mathbf{F} = -\dfrac{\mu m}{R^3} r\, \mathbf{e}_r$)

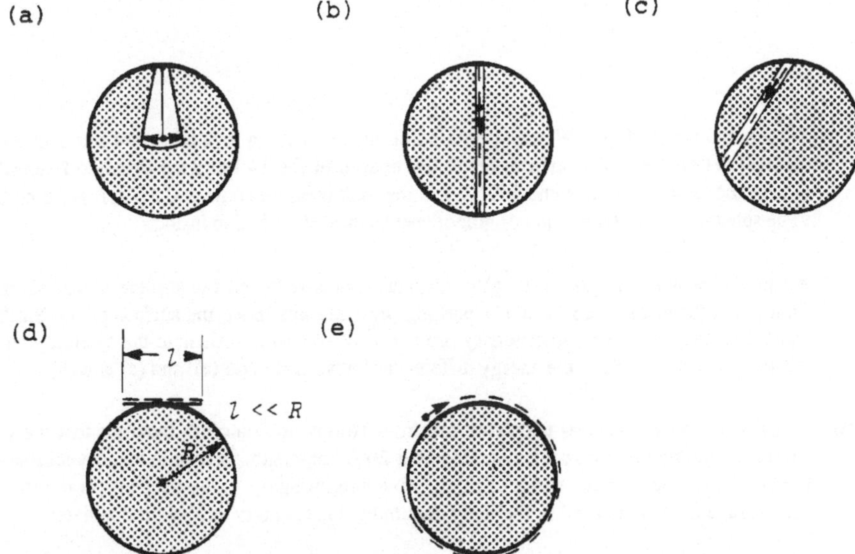

2.7 A mathematical pendulum m of length l is suspended inside a ship which is moving at $v = R\dot{\alpha}$. The angle θ is measured from the local vertical. (a) Write the differential equation of motion for the pendulum, using the angles θ and α, for small vibration amplitudes; (b) Find the pendulum length l for which the pendulum motion is independent of the ship's acceleration $R\ddot{\alpha}$ (= Schuler pendulum); (c) Determine the period of the Schuler pendulum.

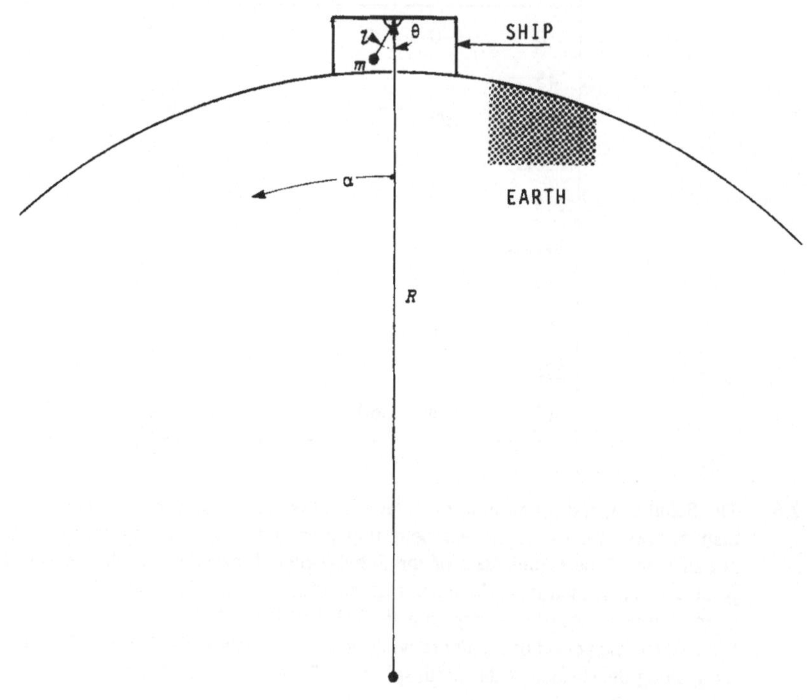

2.8 What is the weight force W acting on a point mass of 100 kg located (a) 14 km above the Earth surface; (b) on the Earth surface; (c) at the bottom of the 14 km deep Erbendorf research shaft (where the temperature is some 300°C, and the rock pressure some 400 MPa)? Assume the Earth to be spherical, nonrotating, and of homogeneous mass distribution inside.

2.9 A satellite of unit mass ($m = 1$ kg) is originally located (a) on the surface of the (nonrotating) Earth, then (b) placed into a circular parking orbit 229 km above the surface of the Earth sphere, and eventually (c) into a geostationary orbit ($r = 42\ 164$ km). Calculate the satellite's energy for (a), (b), and (c), as well as the energy differences between (b) and (a), and (c) and (b).

2.10 A person places an overseas telephone call from Toronto to Frankfurt. Assume that the voice signal travels through ground cables of altogether 2000 km length, then up to a geosynchronous satellite and down again, at the speed of light. How long, roughly, is the pause from the last word of his message to the first word of the reply, assuming that his party answers immediately?

2.11   An orbiter (M = 100 000 kg) circles the Earth at a height of 265 km, with an attitude as shown. A small payload (m = 10 kg) is released from the orbiter's cargo bay with a departure velocity of 0.8 m/s up. Determine the separation between orbiter orbit and payload orbit after a quarter period, if neither payload nor orbiter carry out any orbital manoeuvring after separation.

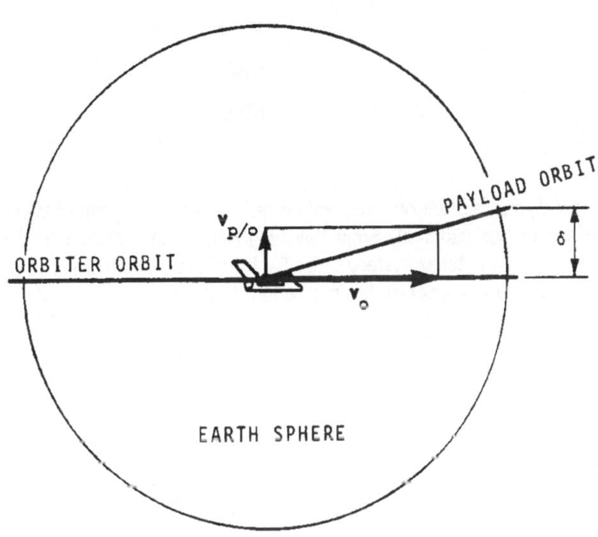

2.12   Show that the eccentricity can be expressed by

$$\varepsilon = \sqrt{1 + \frac{2\,E\,h^2}{\mu^2 m}}$$

2.13   For purposes of the present problem, assume that your own body had a mass of 100 kg, concentrated at the mass centre. Further assume that there is no other mass (Earth, Moon, Sun, or ...) present. A small satellite (m << 100 kg) is circling your body at an orbital radius of 1 m. (a) How long would it take this satellite to complete one orbit? (b) If your mass was only 50 kg, how long would it take the satellite then to complete one orbit?

2.14   Sketch to scale the Earth sphere and four satellite orbits about the Earth. All four have the same perigee point, at 6700 km from Earth centre. Orbit "A" has an eccentricity of 0, orbit "B" has 0.1, orbit "C" has 0.2, and orbit "D" has 0.5. Also complete the following table.

Orbit	r_{PE}	p	b	a
A	6700	6700	6700	6700
B	6700			
C	6700			
D	6700			

2.15   A low-flying satellite on a near-polar orbit around the (spherical) Moon will cover the whole lunar surface in one (sidereal) lunar day. (a) How many orbits will the satellite make for a complete coverage of the lunar surface? (b) By how many degrees will a given (momentary) orbital plane be rotated with respect to the (momentary) orbital plane of the preceding orbit?

2.16   A point satellite of $m_2 = 500$ kg mass on a circular orbit of 7020 km about the Earth is tethered by a (massless) cable to a point satellite of $m_1 = 50$ kg on an orbit of 7000 km radius. (a) Determine the tension force in the cable; (b) What cable diameter is required when Kevlar (permissible stress = 2000 MPa) is used? (c) What is the mass of the cable (Kevlar density = 1440 kg/m^3)?

3

Orbit Insertion

After burnout of the last rocket motor used to launch a spacecraft, the latter moves without any thrust acting. If we also disregard the rather small forces due to solar wind, air resistance, the magnetic field, etc., then the motion of the spacecraft is exclusively determined by the gravitational field of the master mass. At the moment of burnout, the spacecraft is located in the *insertion point* of its orbit. The elements or the orbit are determined by the *position* and the *velocity* of the spacecraft at the moment of insertion into its orbit. We restrict further discussion to a point satellite and also assume that the master mass is of point size. The spacecraft thus moves on a Kepler orbit.

Ideal conditions for practical purposes prevail in case insertion point and orbit periapsis coincide. We shall investigate this case first, together with the mathematically similar case of insertion into the apoapsis. After that, insertion into an arbitrary point of a Kepler orbit will be discussed.

For Earth satellites, insertion into the perigee provides ideal conditions under normal circumstances, because control and guidance problems are minimized. During insertion into the perigee a satellite is closest to the Earth and its flight path is initially horizontal, i.e. parallel to the Earth surface.

For satellites, whose intended orbit is close to Earth, a two- or three-stage rocket burning without interruption is typically used to push the satellite to the insertion point. For satellites, whose insertion point is more than some 300 km above Earth surface, a two-stage ascent is often more economical. After burnout and separation of the first stage, satellite and second-stage rocket coast along the *ascent ellipse*, in whose apogee the second-stage rocket is ignited providing the thrust necessary to insert the satellite into its orbit, whereupon satellite and rocket are separated.

3.1   Insertion into Periapsis (or Apoapsis)

Insertion point and periapsis (or apoapsis) coincide in the present case. In spite of the fact that we assume the master mass to be a point for reasons of theoretical conciseness, we need not ignore that in reality master masses are of finite size and close to spherical in shape. For perfect spheres, we find that insertion into the periapsis or apoapsis means that the velocity vector is horizontal (Figure 3.1), i.e. parallel to the surface of the master sphere.

By means of the speed equation (2.13)

$$v = \sqrt{\mu\left[\frac{2}{r} - \frac{1}{a}\right]} \qquad (3.1)$$

the semi-major axis a of the Kepler orbit can be determined. If the velocity v_o and the position r_o for the insertion point are known, the semi-major axis becomes

$$a = \frac{r_o}{2 - \dfrac{r_o v_o^2}{\mu}} \qquad (3.2)$$

For a given insertion point r_o, let us now consider various insertion velocities. For

$$v_o = 0 \qquad (3.3)$$

we find

$$a = \frac{r_o}{2} \qquad (3.4)$$

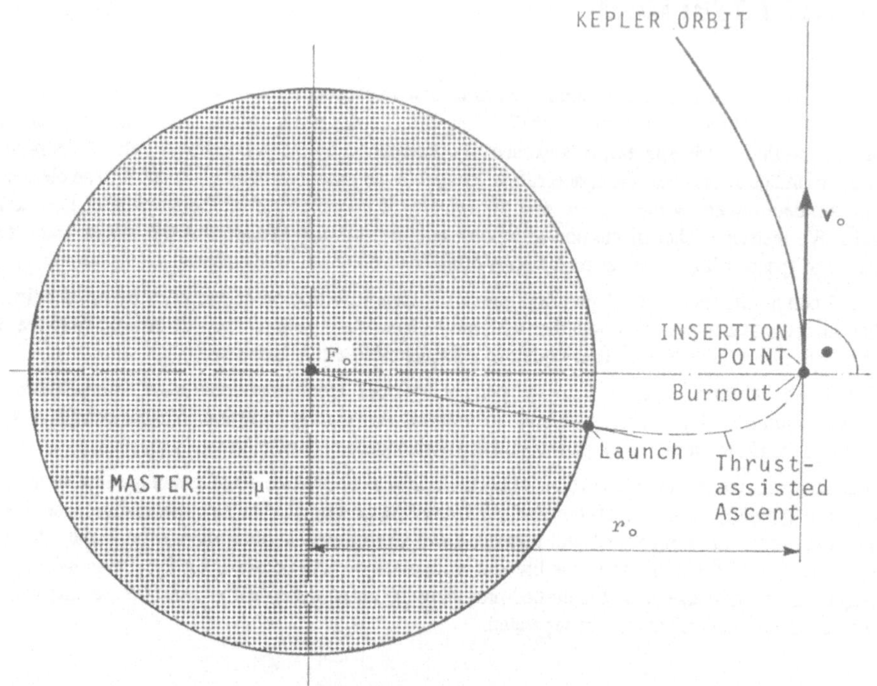

FIGURE 3.1. Insertion of a point satellite into its Kepler orbit at periapsis (or apoapsis).

and the Kepler orbit is a straight line segment (Figure 3.2). For very small insertion velocities

$$0 \; < \; v_o \; < \; \sqrt{\frac{\mu}{r_o}} \tag{3.5}$$

we find

$$\frac{r_o}{2} \; < \; a \; < \; r_o \tag{3.6}$$

representing an ellipse whose apogee coincides with the insertion point. If the insertion velocity is exactly

$$v_o \; = \; \sqrt{\frac{\mu}{r_o}} \tag{3.7}$$

then the orbit becomes a circle of radius

$$a \; = \; r_o \tag{3.8}$$

For insertion velocities below $\sqrt{\mu/r_o}$, insertion point and apoapsis coincide. For insertion velocities in excess of $\sqrt{\mu/r_o}$, insertion point and periapsis coincide.

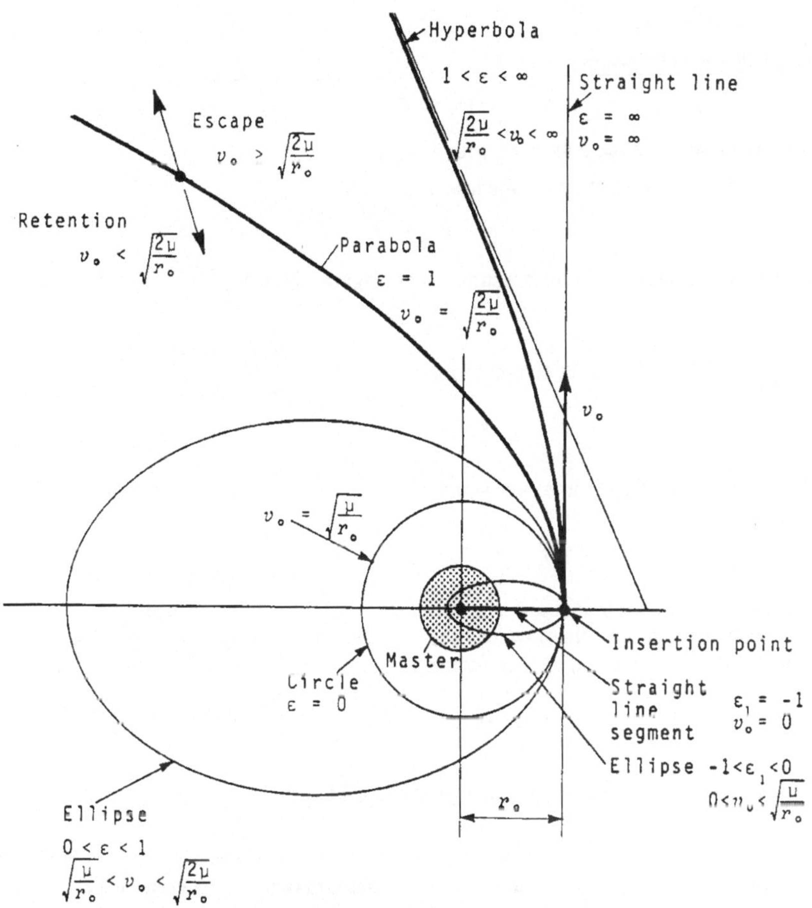

FIGURE 3.2. Insertion into periapsis or apoapsis.

For insertion velocities in the range

$$\sqrt{\frac{\mu}{r_o}} \; < \; v_o \; < \; \sqrt{\frac{2\mu}{r_o}} \tag{3.9}$$

equation (3.2) provides us with

$$r_o \; < \; a \; < \; \infty \tag{3.10}$$

i.e. an elliptic orbit.

For an insertion velocity of exactly

$$v_o \; = \; \sqrt{\frac{2\mu}{r_o}} \tag{3.11}$$

we obtain

$$a \; = \; \infty \tag{3.12}$$

i.e. a parabolic orbit.

For insertion velocities in the range

$$\sqrt{\frac{2\mu}{r_o}} \; < \; v_o \; < \; \infty \tag{3.13}$$

the semi-major axis assumes negative values

$$-\infty \ < \ a \ < \ 0 \tag{3.14}$$

and the satellites move on a hyperbolic path, with $a' = -a$.

A limiting case is represented by an infinite insertion velocity

$$v_o \ = \ \infty \tag{3.15}$$

In this case, the satellite orbit has become a straight line (Figure 3.2), with

$$a = 0 \tag{3.16}$$

The *eccentricity* of the orbits can be computed from

$$\varepsilon_1 \ = \ 1 \ - \ \frac{r_o}{a} \tag{3.17}$$

which, after using equation (3.1), can be expressed as

$$\varepsilon_1 \ = \ \frac{r_o v_o^2}{\mu} \ - \ 1 \tag{3.18}$$

The designation ε_1 has been introduced here because the eccentricity as defined by equations (3.17) or (3.18) can also assume negative values, which occur in cases where *apoapsis* and *insertion point* coincide. However, strict mathematical definition requires the eccentricity, ε, to be positive. We thus have

$$\varepsilon \ = \ |\varepsilon_1| \ = \ \left| \frac{r_o v_o^2}{\mu} \ - \ 1 \right| \tag{3.19}$$

Table 3.1 gives an overview of the various cases of horizontal insertion.

Shape of Orbit	Insertion Velocity v_o	Semi-major Axis a	Eccentricity ε or ε_1	Remarks
Straight Line Segment (free fall)	$v_o = 0$	$a = \dfrac{r_o}{2}$	$\varepsilon_1 = -1$	Insertion into apoapsis
Ellipse	$0 < v_o < \sqrt{\dfrac{\mu}{r_o}}$	$\dfrac{r_o}{2} < a < r_o$	$-1 < \varepsilon_1 < 0$	$r_o = r_{AP}$
Circle	$v_o = \sqrt{\dfrac{\mu}{r_o}}$	$a = r_o$	$\varepsilon = 0$	
Ellipse	$\sqrt{\dfrac{\mu}{r_o}} < v_o < \sqrt{\dfrac{2\mu}{r_o}}$	$r_o < a < \infty$	$0 < \varepsilon < 1$	Insertion into periapsis
Parabola	$v_o = \sqrt{\dfrac{2\mu}{r_o}}$	$a = \pm\infty$	$\varepsilon = 1$	$r_o = r_{PE}$
Hyperbola	$\sqrt{\dfrac{2\mu}{r_o}} < v_o < \infty$	$-\infty < a < 0$ $(0 < a' < \infty)$	$1 < \varepsilon < \infty$	
Straight Line	$v_o = \infty$	$a = 0 \, (a' = 0)$	$\varepsilon = \infty$	

TABLE 3.1. Kepler orbits for insertion into periapsis (or apoapsis).

3.2 Insertion into an Arbitrary Point of the Orbit

We assume that the following quantities are given (Figure 3.3):

$v_o =$ insertion velocity

$r_o =$ distance between insertion point and master mass centre

$\beta_o =$ insertion *flight path angle*, i.e. angular deviation of orbit from the horizontal at the insertion point.

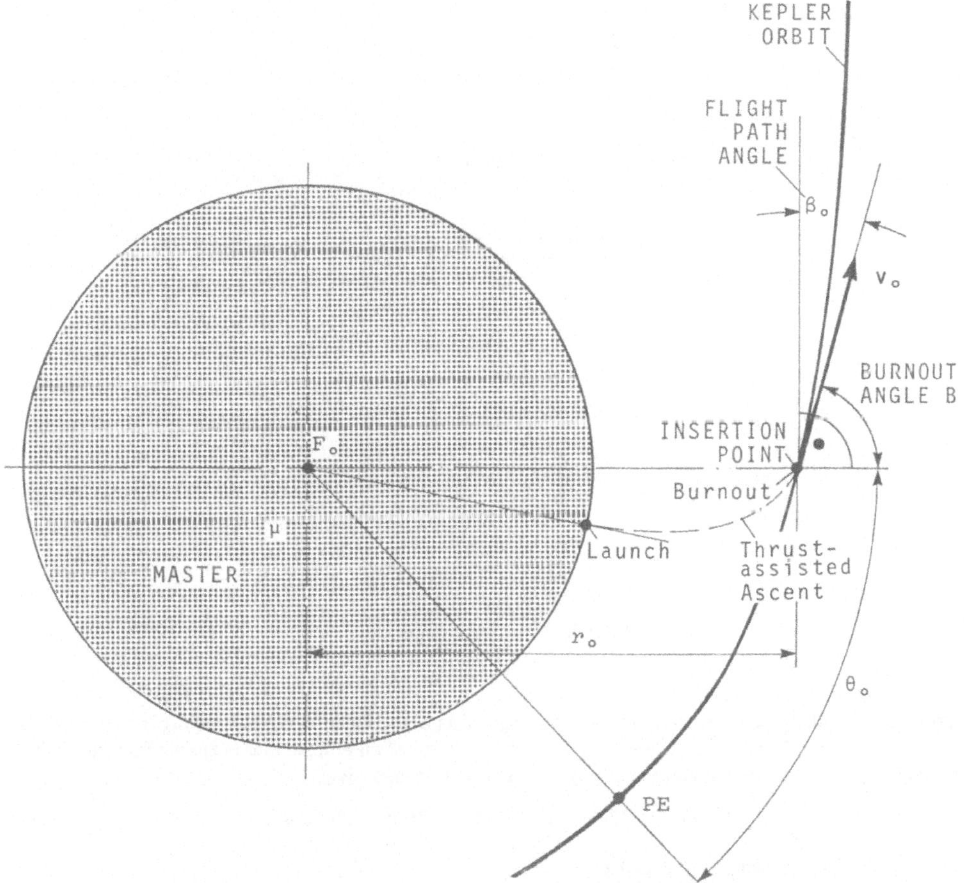

FIGURE 3.3. Insertion of a satellite into its Kepler orbit at an arbitrary point.

The semi-major axis is obtained from equation (3.2),

$$a = \frac{r_o}{2 - \dfrac{r_o v_o^2}{\mu}}$$

The magnitudes of the semi-major axis a as a function of the insertion velocity v_o is plotted in Figure 3.4. From equation (3.2), we conclude that the size of the semi-major axis is independent of the insertion[*] flight path angle β_o.

---

[*] Space engineers refer to B $= 90° - \beta_o$ as the *burnout angle*.

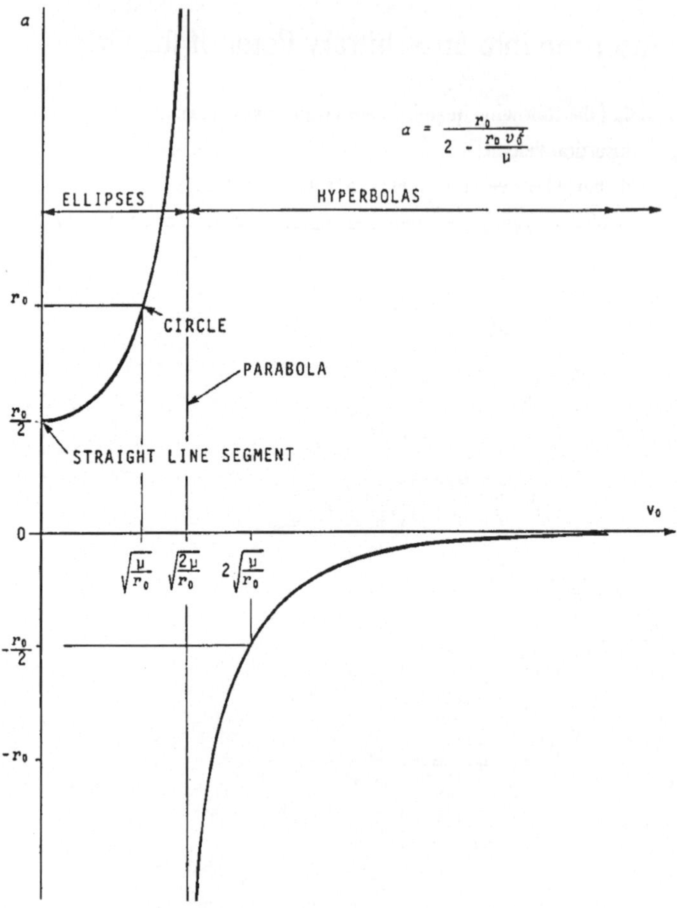

FIGURE 3.4. Semi-major axes.

In Figure 3.5 an ellipse is shown. If the semi-major axis a, the location r_o, and the *flight path angle* β_o, all at insertion, are known and if use is made of the property of an ellipse, that tangent and normal bisect the angle between the lines drawn to each focus, the law of cosines gives

$$(2\varepsilon a)^2 = (2a - r_o)^2 + r_o^2 - 2(2a - r_o)r_o \cos 2\beta_o$$

from which the eccentricity is obtained as

$$\varepsilon_1 = \left[1 - \frac{r_o}{a}\right]\left[1 + \frac{\left[2 - \dfrac{r_o}{a}\right]\dfrac{r_o}{a}}{\left[1 - \dfrac{r_o}{a}\right]^2}\sin^2\beta_o\right]^{\frac{1}{2}} \tag{3.20}$$

We have again used the symbol ε_1, which may also assume negative values.

With the aid of equation (3.2), the eccentricity becomes

$$\varepsilon_1 = \left[\frac{r_o v_o^2}{\mu} - 1\right]\left[1 + \frac{\left[2 - \dfrac{v_o v_o^2}{\mu}\right]\dfrac{r_o v_o^2}{\mu}}{\left[\dfrac{r_o v_o^2}{\mu} - 1\right]^2}\sin^2\beta_o\right]^{\frac{1}{2}} \tag{3.21}$$

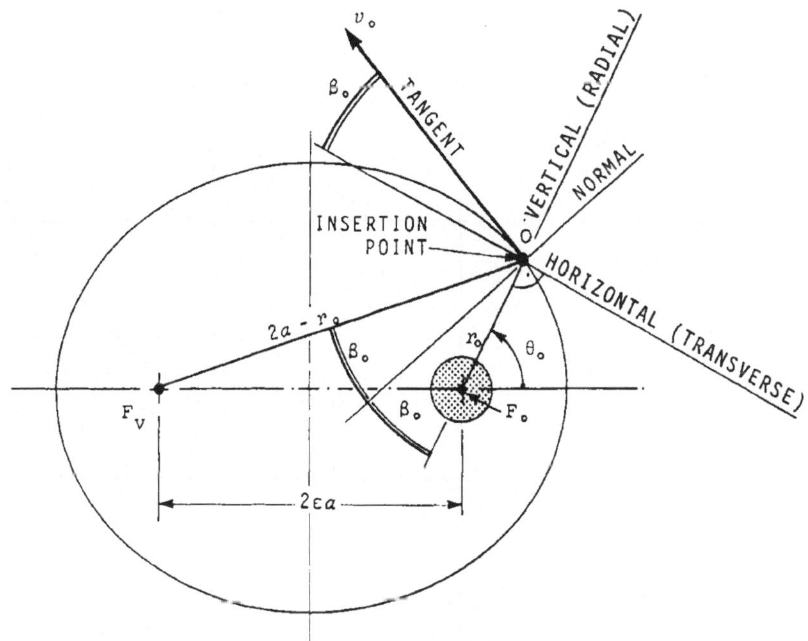

FIGURE 3.5. Insertion point and orbit.

By using $\cos^2\beta_o = (1 + \cos 2\beta_o)/2$, equation (3.21) can also be expressed as

$$\varepsilon = \sqrt{1 - \frac{r_o v_o^2}{2\mu}\left[2 - \frac{r_o v_o^2}{\mu}\right](1 + \cos 2\beta_o)} \tag{3.22}$$

Since the eccentricity, as expressed by equation (3.22) is always positive, the symbol ε can be used.

A plot of eccentricity versus insertion velocity is shown in Figures 3.6 and 3.7. For $v_o = \sqrt{\mu/r_o}$ the eccentricity ε reaches a minimum value given by

$$|\sin \beta_o|$$

Zero eccentricity, i.e. a circular orbit, can only be achieved when $\beta_o = 0°$, i.e. by horizontal insertion.

In order to find the true anomaly θ_o of the insertion point, we inspect Figure 3.5, from which we conclude that

$$(2a - r_o)\sin(\theta_o - 2\beta_o) = r_o \sin\theta_o \tag{3.23}$$

With the aid of a trigonometric identity, we can obtain from equation (3.23)

$$\tan \theta_o = \frac{(2a - r_o)\sin 2\beta_o}{(2a - r_o)\cos 2\beta_o - r_o} \tag{3.24}$$

which, upon introducing equation (3.2), can also be written

$$\tan \theta_o = \frac{\sin 2\beta_o}{\cos 2\beta_o + 1 - \dfrac{2\mu}{r_o v_o^2}} \tag{3.25}$$

The true anomaly θ_o of the insertion point as function of the insertion velocity v_o is plotted in Figure 3.8.

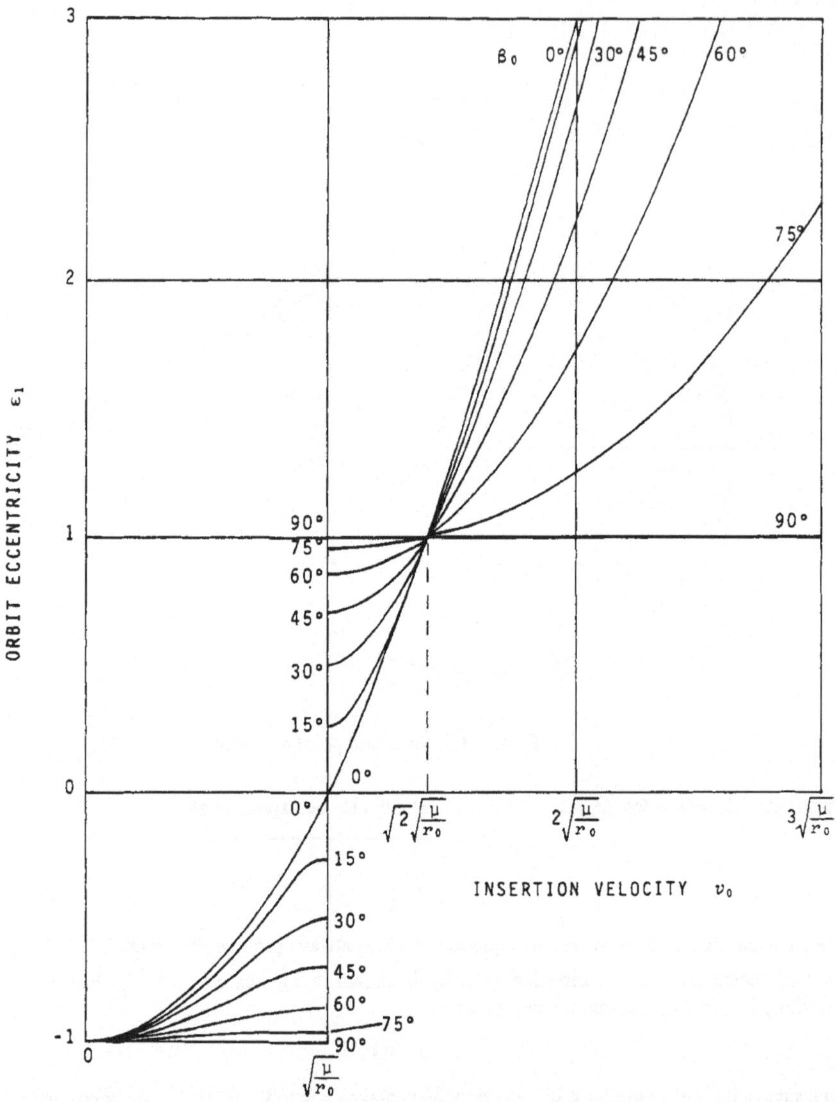

FIGURE 3.6. Orbit eccentricity ε_1.

All of the possible Kepler orbits have the same tangent in the insertion point. Since the tangent in a point on an ellipse bisects the angle between the line to the two foci, we conclude that the vacant foci F_v for any resulting Kepler orbit must lie on one and the same straight line. This line can be easily found, as an inspection of Figure 3.9 will indicate.

Figure 3.10 shows the Kepler orbits which result for a fixed distance r_o of the insertion point from the master, and a fixed flight path angle β_o at the insertion point. Merely the insertion velocity v_o is allowed to assume different values.

A study of Figure 3.10 also shows that the line of apsides, as determined by the location of the two foci F_o and F_v, moves in the direction indicated by the increase of the insertion velocity. A measure of this motion is the true anomaly θ_o of the insertion point, which becomes smaller and smaller as the insertion velocity becomes larger and larger.

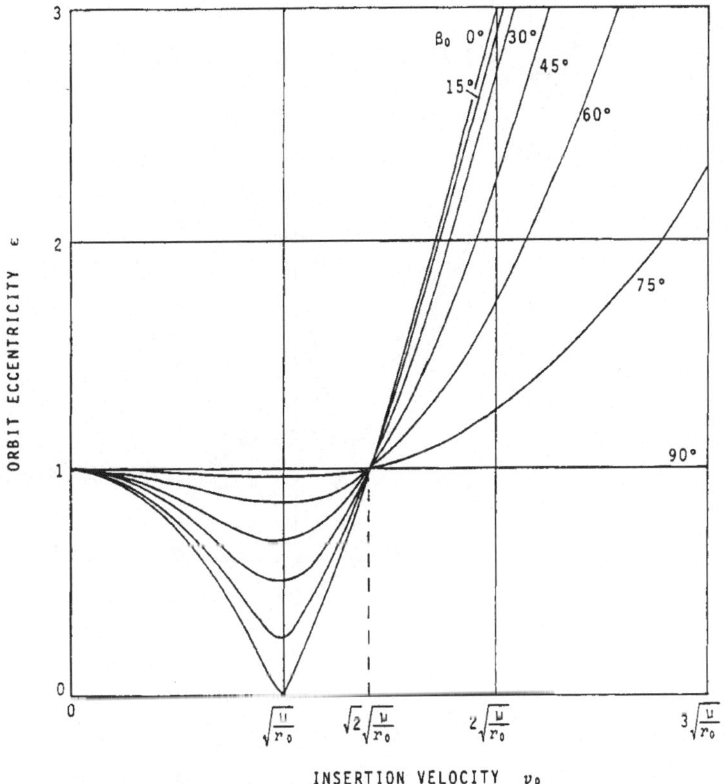

FIGURE 3.7. Orbit eccentricity ε.

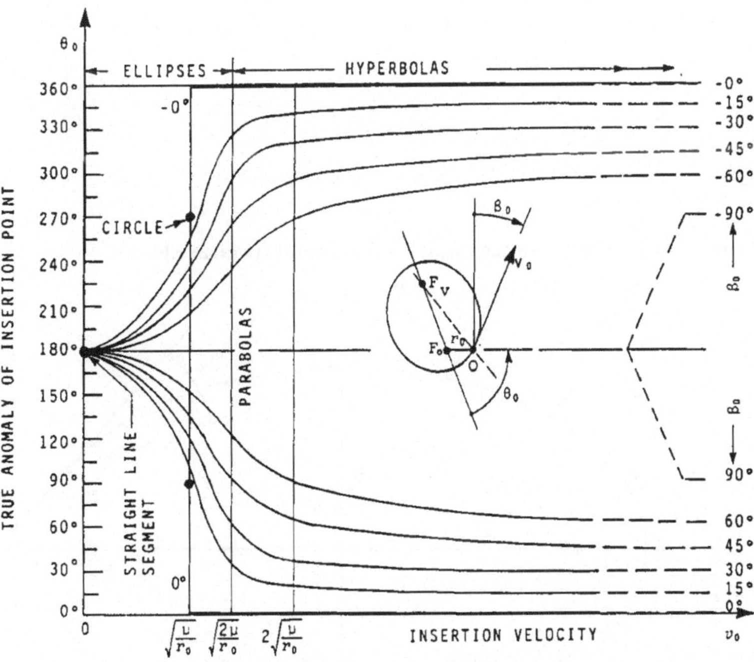

FIGURE 3.8. True anomaly of insertion point.

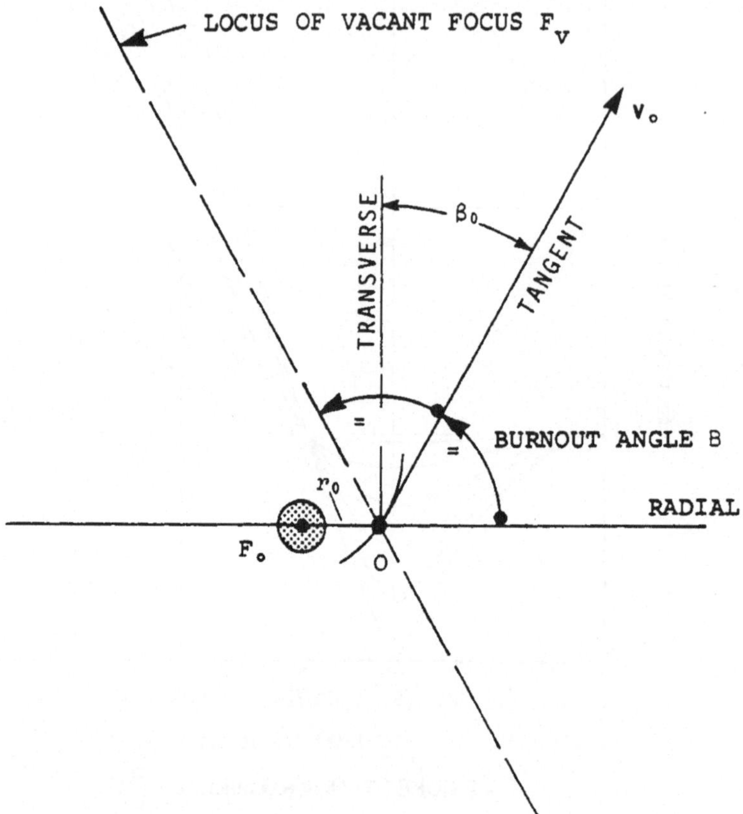

FIGURE 3.9. Insertion point and locus of vacant focus.

Problems

3.1   Given is r, a, β_o. Derive an expression for the eccentricity of an ellipse

$$\varepsilon = f(r, a, \beta_o) = ?$$

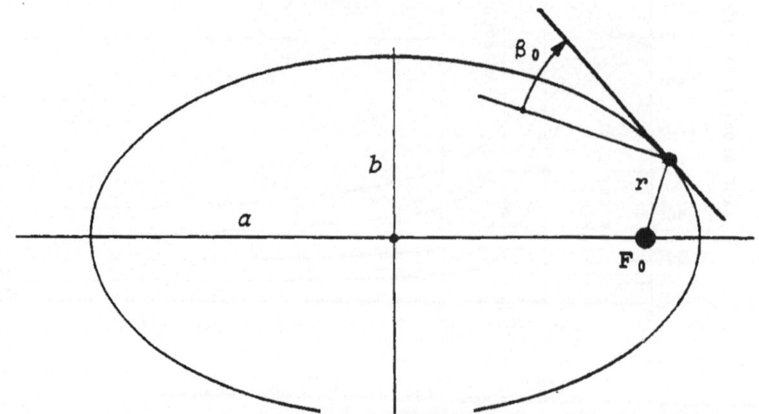

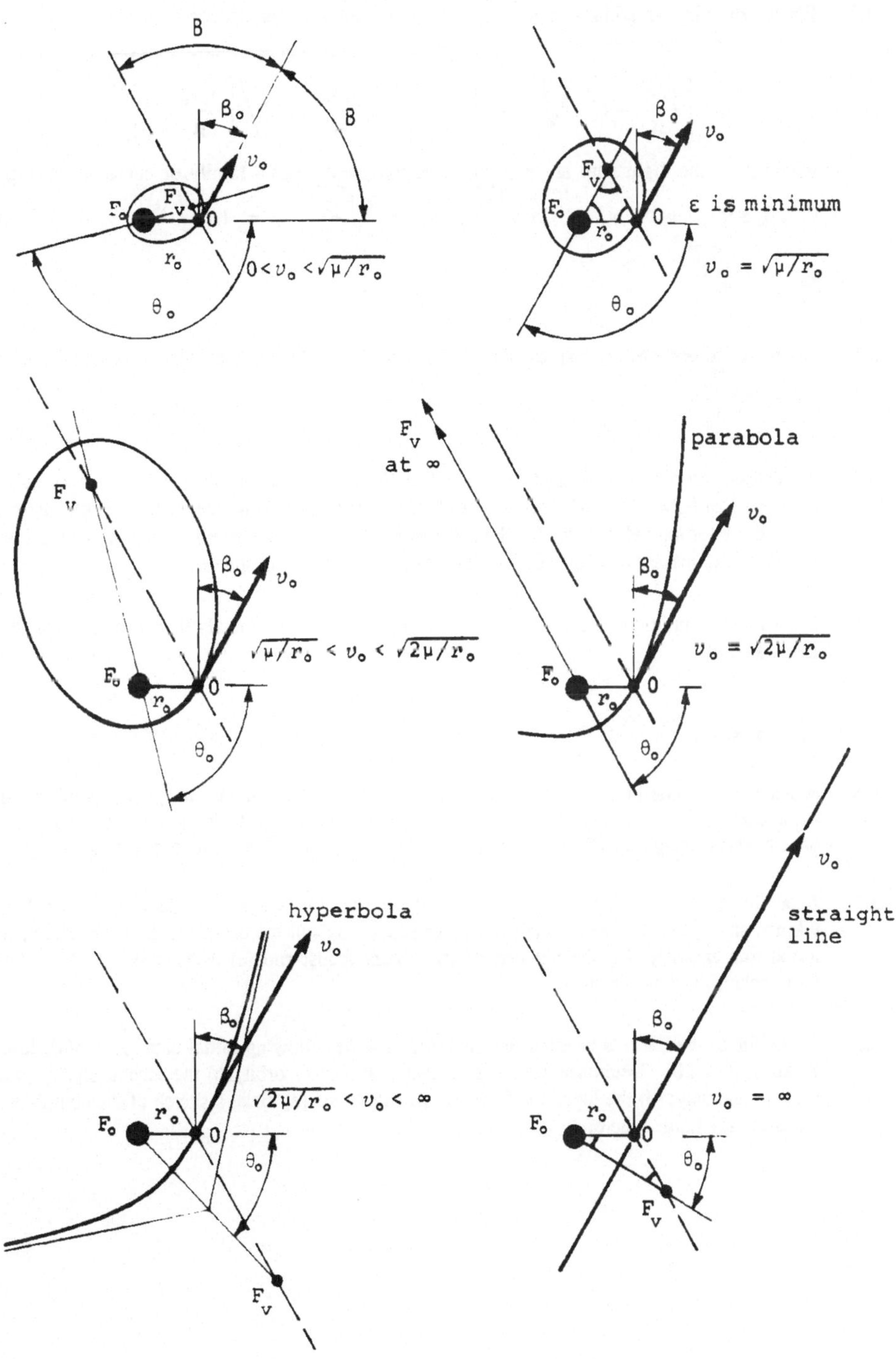

FIGURE 3.10. Orbits resulting from various insertion speeds.

3.2 For insertion into an arbitrary point (r_o, θ_o) of the orbit, the orbit eccentricity is given by

$$\varepsilon_1 = \left[\frac{r_o v_o^2}{\mu} - 1 \right] \sqrt{ 1 + \left[2 - \frac{r_o v_o^2}{\mu} \right] \frac{r_o v_o^2}{\mu} \sin^2\beta_o \Bigg/ \left[\frac{r_o v_o^2}{\mu} - 1 \right]^2 }$$

where β_o is the flight path angle in the insertion point. Plot a family of curves $\varepsilon = \varepsilon(\beta_o)$ for $0° \leq \beta_o \leq 90°$, with $v_o = 0.9\sqrt{\dfrac{\mu}{r_o}}$, $v_o = \sqrt{\dfrac{\mu}{r_o}}$, $v_o = 1.1\sqrt{\dfrac{\mu}{r_o}}$, $v_o = \sqrt{2}\sqrt{\dfrac{\mu}{r_o}}$, $v_o = 2\sqrt{\dfrac{\mu}{r_o}}$.

3.3 Given is the eccentricity expression of Problem 3.2. What is the value assumed by ε_1 when $v_o = \sqrt{\dfrac{\mu}{r_o}}$?

3.4 A satellite, inserted into its orbit at a height of 629 km above the Earth sphere surface at a flight path angle of $\beta_o = 10°$, is to assume a minimum eccentricity orbit. Determine (a) the eccentricity ε; (b) the insertion velocity v_o; (c) the semi-major axis a; (d) the true anomaly θ_o of the insertion point; and (e) make a careful sketch of the orbit and of the Earth sphere.

3.5 Show that for minimum eccentricity orbit, insertion point true anomaly θ_o and flight path angle β_o are related by

$$\theta_o = 90° + \beta_o$$

Make a sketch.

3.6 A satellite is inserted into its orbit at a height of 1629 km above the Earth sphere's surface. It is to assume an orbit of minimum eccentricity, and to remain safely outside the Earth's atmosphere (i.e. have a perigee height of 200 km). Determine the flight path angle at the insertion point.

3.7 A satellite of the Earth is inserted into its orbit, with the following initial data: $r_o = 8000$ km, $v_o = 9$ km/s, and $\beta_o = 30°$. Determine (a) the semi-major axis of the orbit; (b) the eccentricity; (c) the initial true anomaly; (d) the location of the vacant focus; and (e) make a sketch (including the Earth sphere) and comment!

3.8 A satellite of the Earth is inserted into its orbit with the following initial data: $r_o = 8000$ km, $v_o = 10$ km/s, $\beta_o = 20°$. Determine (a) the semi-major axis of its orbit; (b) the eccentricity; (c) the initial true anomaly; (d) the location of the vacant focus; and (e) make a sketch of the complete orbit, including the Earth sphere.

4

Transfer

The manoeuvre where a satellite is to be moved from one orbit to another is called *transfer*. If original and final orbit intersect one another, the transfer can take place at the point of intersection, which is then called the *transfer point*. If original and final orbit do not intersect, then transfer can only take place along an intermediate orbit, called the *transfer orbit*. The transfer of a satellite from one orbit to another is a manoeuvre that is carried out frequently, since satellites can only rarely be inserted into their final orbit immediately after launching.

Transfers of the first kind are those concerned with moves from original to final orbit *about the same master*. These transfers are the subject of the present chapter. Transfers of the second kind involve an original orbit about one master, and a final orbit about *a new master*. Such transfers are dealt with in a separate chapter, on Satellites from Infinity.

Minor orbit errors in position, height, velocity (magnitude and direction), can usually not be prevented when a satellite is first inserted into an orbit. By means of a transfer manoeuvre, it is possible to effect orbit corrections bringing one or several orbit elements to within the limits prescribed for the mission.

If the mission profile provides for a temporary parking of a satellite on a parking orbit, then the subsequent insertion into the final orbit again represents a transfer.

In the case of *rendezvous* of two satellites, the initial orbit of the satellite to be transferred, and the orbit of the target satellite are generally known. The orbit of the target satellite is the final orbit for the transfer satellite. Not only original and final orbit are of significance, but also *time* of arrival and *point* of arrival on the target orbit. The departure of the transfer satellite from its original orbit must occur at a time and place ("launch window"), that assure arrival at the final orbit at the moment, where the target vehicle passes through the point of intersection of its own orbit with that of the transfer orbit of the intercepting vehicle.

4.1   Single Impulse

Let us now investigate, what can be achieved by a single short impulse, causing a change of velocity which we shall refer to as a *velocity kick*. If the impulse is of sufficiently short duration, one can justify the assumption that the satellite is located at *one* point on its orbit during the action of the impulse. This point is the transfer point T (Figure 4.1).

Velocity kicks Δv are typically brought about by small on-board rockets, supplying a burst of thrust F for a short period of time Δt. The principle of impulse and momentum states that

$$\mathbf{F}\, dt = d(m\mathbf{v})$$

For practical purposes, it can usually be assumed that the rocket thrust F is constant, and that the amount of rocket fuel used during a thrust burst is so small, that the satellite mass m can be considered constant as well. We thus conclude, first, that the direction of the velocity kick Δv is equal to the direction of the rocket thrust F, and second, that the magnitude Δv of the velocity kick is directly proportional to the duration Δt of the rocket burn,

$$\Delta \mathbf{v} = \frac{\mathbf{F}}{m}\, \Delta t$$

Using orbit coordinates of the original orbit, the velocity kick becomes

$$\Delta \mathbf{v} = [\mathbf{e}_r\, \mathbf{e}_\theta\, \mathbf{e}_z] \begin{bmatrix} \Delta v_r \\ \Delta v_\theta \\ \Delta v_z \end{bmatrix} \tag{4.1}$$

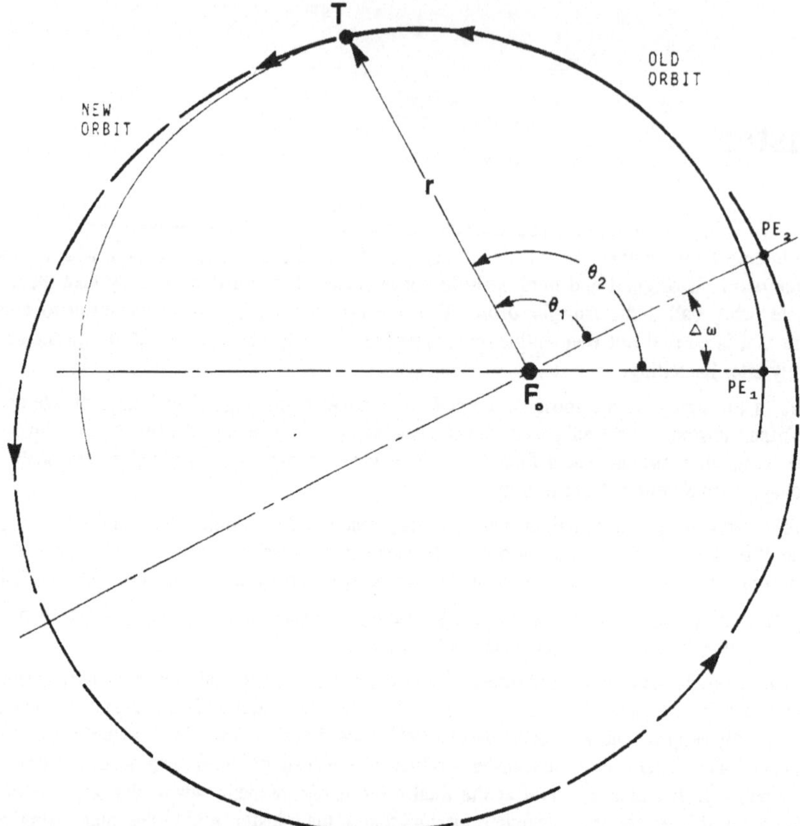

FIGURE 4.1. Transfer in point T.

We will now investigate the influence of the components of the velocity kick upon the geometric orbit elements, and shall later derive the conditions under which the transfer into a new orbit with prescribed orbit elements can be achieved by means of a single impulse.

4.2   Coplanar Transfer

If at an arbitrary position **r** a satellite is subjected to a velocity kick, which does not have a component normal to the orbital plane, then original orbit and final orbit lie in one and the same plane; they are *coplanar*. We designate quantities of the original orbit by the subscript 1 and those of the new orbit by 2 (Figure 4.2). The velocity kick itself is

$$\Delta \mathbf{v} = [\mathbf{e}_r \ \mathbf{e}_\theta \ \mathbf{e}_z] \begin{bmatrix} \Delta v_r \\ \Delta v_\theta \\ 0 \end{bmatrix} \tag{4.2}$$

The velocity kick takes place at the position r = constant. The transverse velocity component changes from v_{θ_1} to $v_{\theta_1} + \Delta v_\theta$, the radial component from v_{r_1} to $v_{r_1} + \Delta v_r$. As we deal with one and the same master, the radial direction and the transverse direction remain the same. Since the transfer takes place in one and the same plane, only the following orbital elements will be affected: the semi-major axis, the eccentricity and the position of the periapsis.

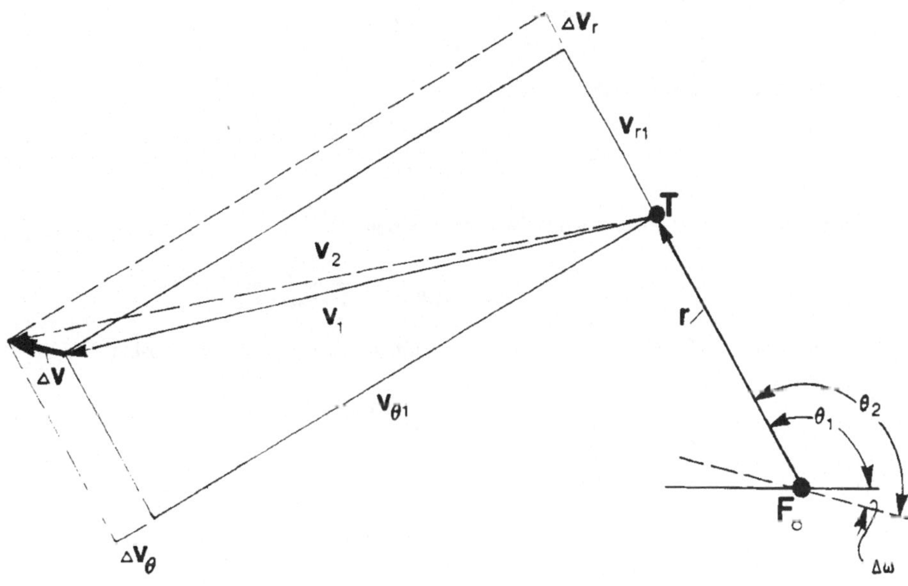

FIGURE 4.2. Velocity kick Δv.

The *semi-major axis* of the new orbit is obtained from the speed equation (2.13) solved for the position.

$$r \;=\; \frac{2a_1}{1 + \dfrac{a_1 v_1^2}{\mu}} \;=\; \frac{2a_2}{1 + \dfrac{a_2 v_2^2}{\mu}} \tag{4.3}$$

Considering that

$$v_2 \;=\; \sqrt{\left[v_{r_1} + \Delta v_r \right]^2 \;+\; \left[v_{\theta_1} + \Delta v_\theta \right]^2} \tag{4.4}$$

one obtains

$$a_2 \;=\; \frac{a_1}{1 - \dfrac{a_1}{\mu}\left(v_2^2 - v_1^2 \right)}$$

or

$$a_2 \;=\; \frac{a_1}{1 - \dfrac{a_1}{\mu}\left[2v_{r_1}\Delta v_r + 2v_{\theta_1}\Delta v_\theta + \Delta v_r^2 + \Delta v_\theta^2 \right]} \tag{4.5}$$

The specific angular momentum of the satellite on the original orbit is

$$h_1 \;=\; r\, v_{\theta_1} \tag{4.6}$$

On the new orbit, it is

$$h_2 = rv_{\theta_2} = h_1 + r\Delta v_\theta \tag{4.7}$$

Using

$$h = \sqrt{\mu p} = \sqrt{\mu a(1-\varepsilon^2)} \tag{4.8}$$

we obtain for the *eccentiricity* of the new orbit

$$\varepsilon_2 = \sqrt{1 - \frac{h_2^2}{\mu a_2}} \tag{4.9}$$

Finally we employ the polar equation for a conic section

$$r = \frac{p_1}{1+\varepsilon_1\cos\theta_1} = \frac{p_2}{1+\varepsilon_2\cos\theta_2} \tag{4.10}$$

For the *true anomaly* θ_2 of the satellite on its new orbit, the transcendental relationship

$$\cos\theta_2 = \frac{1}{\varepsilon_2}\left[\frac{p_2}{p_1}(1+\varepsilon_1\cos\theta_1) - 1\right] \tag{4.11}$$

or

$$\cos\theta_2 = \frac{1}{\varepsilon_2}\left[\frac{p_2}{r} - 1\right] \tag{4.12}$$

is obtained. The *argument of periapsis* of the new orbit is moved by the amount

$$\Delta\omega = \theta_1 - \theta_2 \tag{4.13}$$

in relation to the original periapsis position (Figure 4.1).

By means of equation (4.5), (4.9) and (4.13) the elements of the new Kepler orbit can be computed.

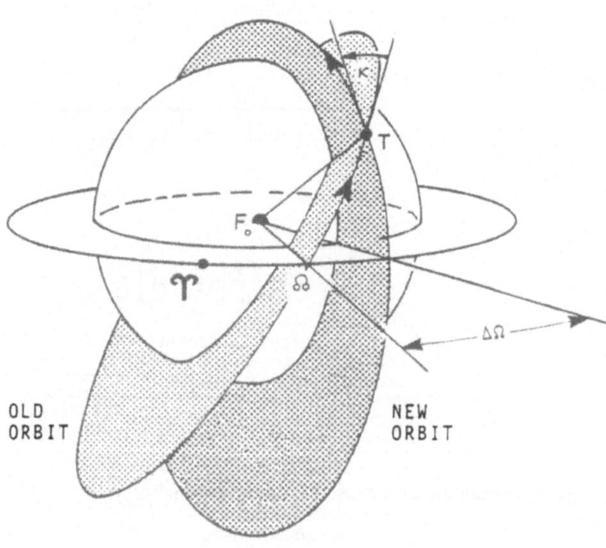

FIGURE 4.3. Tilting of the orbital plane.

4.3 Change of Orbital Plane

In the preceding article we considered a single velocity kick within the orbital plane, which led to a change in size and shape of orbit, not however to a change of orientation of the orbital plane. To change the orientation of the orbital plane in space requires a velocity kick with a component normal to the original plane. A *pure orbital plane change* - without associated size or shape changes of the orbit - requires that only the *direction* of the transverse velocity component is changed by the velocity kick (Figure 4.3), while its magnitude and the magnitude and direction of the radial velocity component remain unchanged. From the isosceles velocity triangle of Figure 4.4, we obtain for the tilt angle κ

$$\sin\frac{\kappa}{2} \;=\; \frac{\Delta v}{2v_\theta} \tag{4.14}$$

with v_θ = constant. Using the coordinate system for the old orbit, the velocity kick required for a simple change of the orientation of the orbital plane must have the following components

$$\Delta v \;=\; [e_r,\, e_\theta\, e_z]
\begin{bmatrix}
0 \\[6pt]
-2v_\theta \sin^2\dfrac{\kappa}{2} \\[6pt]
2v_\theta \sin\dfrac{\kappa}{2}\cos\dfrac{\kappa}{2}
\end{bmatrix} \tag{4.15}$$

or

$$\Delta v \;=\; [e_r,\, e_\theta\, e_z]
\begin{bmatrix}
0 \\[6pt]
-v_\theta(1-\cos\kappa) \\[6pt]
v_\theta \sin\kappa
\end{bmatrix}$$

The axis about which the orbital plane is tilting coincides with the position vector r of the satellite (Figure 4.4) and represents the line of intersection between old and new orbital plane.

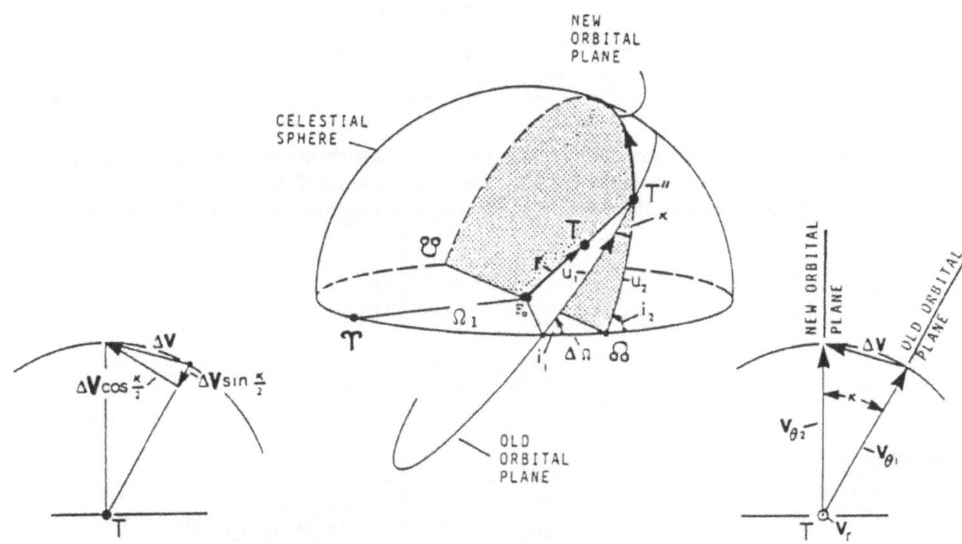

FIGURE 4.4. Tilt angle and velocity kick.

Let us assume that the inclination angle i_1 and the orbit angle u_1 on the old orbit are known. By means of the cosine law for plane angles of spherical trigonometry, the inclination angle i_2 of the new orbital plane can be found.

$$\cos i_2 \;=\; \cos i_1 \cos\kappa \;-\; \sin i_1 \sin\kappa \cos u_1 \tag{4.16}$$

The change $\Delta\Omega$ of the right ascension of the ascending node can be found from

$$\cot\Delta\Omega \;=\; \frac{\sin i_1 \cot\kappa + \cos i_1 \cos u_1}{\sin u_1} \tag{4.17}$$

The new orbit angle u_2 can then be obtained from

$$\cot u_2 \;=\; \frac{\sin\kappa \cot i_1 + \cos\kappa \cos u_1}{\sin u_1} \tag{4.18}$$

4.4 Transfer to a Prescribed Target Orbit

In practical cases, the problem is often to determine the velocity kick required for a transfer, when the elements of original and final orbit are given. The transfer by means of a single impulse is, of course, only possible when original and final orbit have a point of intersection, which then also becomes the point in which the transfer has to be effected. Original and final orbit intersect in the transfer point T. Since all orbital elements of the old and new orbits are known, one can establish the direction of the line of intersection of the two orbital planes and thus also the transfer point T.

$$\mathbf{e}_{rT} \;=\; \frac{\mathbf{e}_{n_1}\times\mathbf{e}_{n_2}}{\left|\mathbf{e}_{n_1}\times\mathbf{e}_{n_2}\right|} \tag{4.19}$$

where $\mathbf{e}_{rT}$ is the unit vector pointing towards the transfer point T, and $\mathbf{e}_{n_1}$ and $\mathbf{e}_{n_2}$ are unit vectors normal to the old and new orbital planes.

The transverse directions at the transfer point are

$$\mathbf{e}_{\theta_1} \;=\; \frac{\mathbf{e}_{n_1}\times\mathbf{e}_{rT}}{\left|\mathbf{e}_{n_1}\times\mathbf{e}_{rT}\right|} \tag{4.20}$$

$$\mathbf{e}_{\theta_2} \;=\; \frac{\mathbf{e}_{n_2}\times\mathbf{e}_{rT}}{\left|\mathbf{e}_{n_2}\times\mathbf{e}_{rT}\right|} \tag{4.21}$$

As an example, take two circular orbits of 7000 km radius each about the Earth. Orbit 1 is in the equatorial plane, orbit 2 has an inclination of 20°. The line of intersection of the two orbits extends from Earth centre to the first point of Aries. Using $OXYZ$ coordinates with X pointing to Aries and Z pointing North,

$$\mathbf{e}_{n_1} \;=\; [\mathbf{e}_X\;\mathbf{e}_Y\;\mathbf{e}_Z]\begin{bmatrix}0\\0\\1\end{bmatrix}$$

$$\mathbf{e}_{n_2} \;=\; [\mathbf{e}_X\;\mathbf{e}_Y\;\mathbf{e}_Z]\begin{bmatrix}0\\-\sin 20°\\\cos 20°\end{bmatrix}$$

such that

$$\mathbf{e}_{rT} \;=\; \frac{1}{\left|\mathbf{e}_{n_1}\times\mathbf{e}_{n_2}\right|}\begin{vmatrix}\mathbf{e}_X & \mathbf{e}_Y & \mathbf{e}_Z\\0 & 0 & 1\\0 & -\sin 20° & \cos 20°\end{vmatrix} \;=\; [\mathbf{e}_X\;\mathbf{e}_Y\;\mathbf{e}_Z]\begin{bmatrix}1\\0\\0\end{bmatrix}$$

and

$$e_{\theta_2} = \frac{1}{|e_{n_2} \times e_{rT}|} \begin{vmatrix} e_X & e_Y & e_Z \\ 0 & -\sin 20^\circ & \cos 20^\circ \\ 1 & 0 & 0 \end{vmatrix} = [e_X \ e_Y \ e_Z] \begin{bmatrix} 0 \\ \cos 20^\circ \\ \sin 20^\circ \end{bmatrix}$$

Since the orbital elements of both orbits are all assumed known, the periapses r_{PE_1} and r_{PE_2} are also known. The true anomalies of the transfer point on the old and the new orbit can be obtained from

$$\cos\theta_1 = \frac{r_{PE_1} \cdot e_{rT}}{r_{PE_1}} \tag{4.22}$$

and

$$\cos\theta_2 = \frac{r_{PE_2} \cdot e_{rT}}{r_{PE_2}} \tag{4.23}$$

When the true anomalies of the transfer point are known, the velocities v_1 and v_2 of the satellite in the transfer point, immediately prior to transfer on the old orbit, and immediately after transfer on the new orbit, can be determined. The velocity kick required for the transfer is then

$$\Delta v = v_2 - v_1 \tag{4.24}$$

If old and new orbit have no common point of intersection, then at least two impulses are required to effect a transfer.

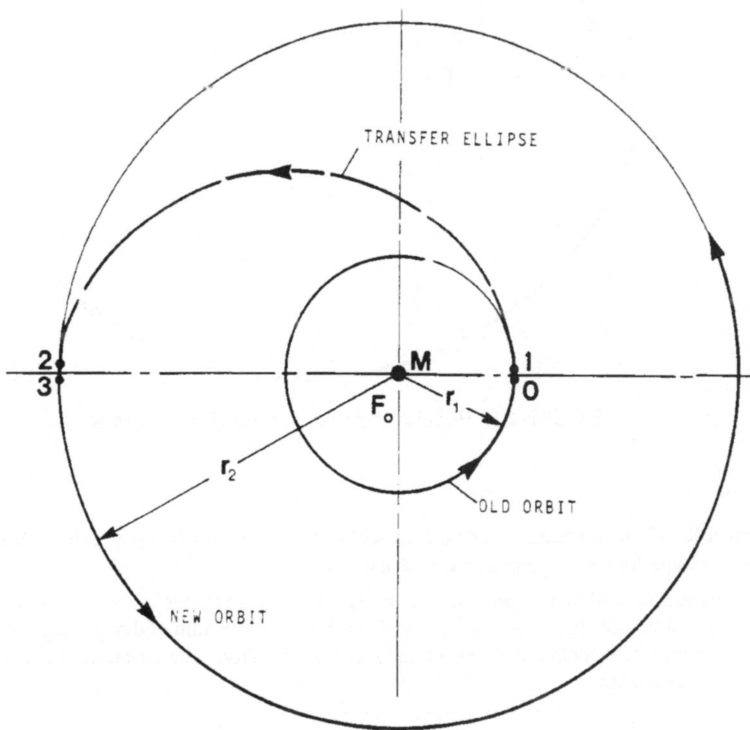

FIGURE 4.5. Hohmann transfer.

4.5   Hohmann Transfer

Hohmann[*] (1925) in his early work already described the transfer between two coplanar and concentric circular orbits (Figure 4.5). For a Hohmann transfer, *two* velocity kicks in transverse direction are required. The first pushes the satellite onto the *transfer ellipse*, the second is applied at the moment where the satellite reaches the apoapsis of the transfer ellipse. It inserts the satellite into its final orbit. Hohmann proposed the application of velocity kicks without radial components. As was shown later (Ting, 1960) Hohmann transfers represent transfers for which the required energy (and thus fuel consumption) is a minimum. This applies not only in cases of a transfer between circular orbits, but also for transfers between elliptical orbits.

In order to effect a Hohmann transfer between ellipses, the major axes of the old and new ellipses must coincide, the periapses of the transfer orbit and the smaller orbit must coincide, and the periapsis of the larger orbit and the apoapsis of the transfer orbit must coincide (Figure 4.6).

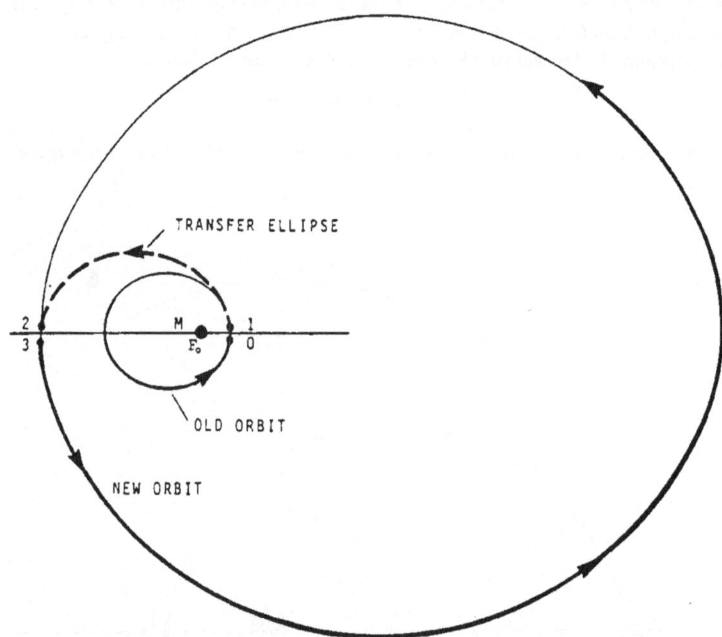

FIGURE 4.6. Hohmann transfer between elliptical orbits.

The sum of the absolute values of the velocity kicks required for a transfer is called *characteristic velocity*. It is a measure for the fuel required for the transfer.

In the following, we will investigate a Hohmann transfer from a smaller to a larger circular orbit. For the inverse process of transfer from a larger to a smaller orbit, the same train of thought applies. The two velocity kicks will then be negative and cause a reduction of velocity. The characteristic velocity, however, remains a positive quantity.

---

[*]   Walter Hohmann (1880-1945), German engineer.

4.6 Hohmann Transfer onto Larger Circular Orbit

Let us now investigate the Hohmann transfer of a satellite from a circular orbit of radius r_1 to a larger, coplanar, concentric orbit of radius r_2 (Figure 4.5). The semi-major axis of the transfer ellipse is then

$$a = \frac{1}{2}(r_1 + r_2) \tag{4.25}$$

For the velocity of the satellite on the transfer ellipse we get

$$v^2 = \mu\left[\frac{2}{r} - \frac{1}{a}\right] = 2\mu\left[\frac{1}{r} - \frac{1}{r_1 + r_2}\right] \tag{4.26}$$

In particular, the velocity in the periapsis r_1 is

$$v_1 = \sqrt{\frac{\mu}{r_1}} \sqrt{\frac{2}{1 + \dfrac{r_1}{r_2}}} \tag{4.27}$$

The velocity of the satellite on the original orbit is

$$v_0 = \sqrt{\frac{\mu}{r_1}} \tag{4.28}$$

The velocity kick Δv_1 which must be applied to the satellite to change its velocity from v_0 to v_1 is

$$\Delta v_1 = v_1 - v_0 = \sqrt{\frac{\mu}{r_1}}\left[\sqrt{\frac{2}{1 + \dfrac{r_1}{r_2}}} - 1\right] \tag{4.29}$$

In the apoapsis r_2 of the transfer ellipse, the velocity (4.26) of the satellite is

$$v_2 = \sqrt{\frac{\mu}{r_2}} \sqrt{\frac{2}{1 + \dfrac{r_2}{r_1}}} \tag{4.30}$$

The velocity which the satellite must have in order to continue its flight on the final circular orbit is

$$v_3 = \sqrt{\frac{\mu}{r_2}} \tag{4.31}$$

The velocity kick required in the second transfer point is thus

$$\Delta v_2 = v_3 - v_2 = \sqrt{\frac{\mu}{r_2}}\left[1 - \sqrt{\frac{2}{1 + \dfrac{r_2}{r_1}}}\right] \tag{4.32}$$

The sum of the two velocity kicks is the *characteristic velocity*.

$$\Delta v = \Delta v_1 + \Delta v_2 \tag{4.33}$$

With the aid of equations (4.5) and (4.8), the characteristic velocity becomes

$$\Delta v = \sqrt{\frac{\mu}{r_1}}\left\{\sqrt{\frac{2}{1 + r_1/r_2}}\left[1 - \frac{r_1}{r_2}\right] + \sqrt{\frac{r_1}{r_2}} - 1\right\} \tag{4.34}$$

Using equation (4.34) for the characteristic velocity of a Hohmann transfer, the following special cases arise:

1)   For $r_2 = r_1$, there is no transfer. The satellite remains on its original circular orbit. The transfer ellipse is reduced to a semi-circle of radius r_1. As is to be expected, we find

$$\Delta v = 0 \tag{4.35}$$

2)   The curve $\Delta v = f(r_2/r_1)$ contains a maximum, located at that value of r_2/r_1 for which

$$\frac{d\Delta v}{d\left[r_2/r_1\right]} = 0 \tag{4.36}$$

Differentiating equation (4.34) with respect to (r_2/r_1) and equating the derivative to zero, shows that the maximum of Δv occurs at

$$\frac{r_2}{r_1} = 15.5817 \tag{4.37}$$

and is

$$\Delta v_{max} = 0.536\, v_o \tag{4.38}$$

3)   If $r_2 = \infty$, then we find the apoapsis of the transfer ellipse at infinity. The transfer ellipse has degenerated into a parabola. The satellite escapes. Only *one* velocity kick is applied. The second velocity kick equals zero, $\Delta v_2 = 0$.

$$\Delta v = \Delta v_1 + 0 = \sqrt{\frac{\mu}{r_1}}\, (\sqrt{2} - 1) = 0.414\, v_0 \tag{4.39}$$

The characteristic velocity required to have the satellite escape is thus smaller than the one required to place a satellite on the finite orbit described in case 2. A closer scrutiny of the behaviour of equation (4.34) will indicate, that when

$$3.304 < \frac{r_2}{r_1} < \infty \tag{4.40}$$

a larger characteristic velocity is required than that which would send the satellite to infinity.

4.7   Energy Increase Due to Velocity Kick

Assuming that the mass of a satellite does not change, while a velocity kick occurs, the *kinetic energy* changes, as the velocity is increased from v to $v + \Delta v$, by an amount

$$\Delta T = \frac{m}{2}(v + \Delta v)^2 - \frac{m}{2}v^2 \tag{4.41}$$

or

$$\Delta T = mv\,\Delta v + \frac{m(\Delta v)^2}{2} \tag{4.42}$$

For most transfer manoeuvres carried out in practice, we have $\Delta v \ll v$, which justifies neglecting the second term on the right hand side of equation (4.42). We consequently obtain for the kinetic energy change

$$\Delta T = m\, v\, \Delta v \tag{4.43}$$

and observe that for small velocity kicks, *the energy increase ΔT is directly proportional to the velocity kick Δv*. In addition, the energy increase is also directly proportional to the satellite velocity v itself. Since the velocity kick Δv is a function of the impulse (thrust times burn duration of the rocket burst) applied to the satellite (and independent of the velocity of the satellite), we conclude that *the energy increase is the greater, the greater the velocity of the satellite.*

The satellite assumes its greatest velocity in the periapsis of its orbit. We thus find: *For a given velocity kick, the increase of kinetic energy is the larger, the closer the satellite is to its master.*

4.8   Fuel Requirements

The thrust F produced by a rocket motor is

$$F = \dot{m}_f c \tag{4.44}$$

where $\dot{m}_f$ is the mass flow of the rocket fuel leaving, and c is the exhaust speed.

If a spacecraft of mass m is given a velocity kick Δv by a thrust F during a short time interval Δt, then

$$F\Delta t = m\Delta v \tag{4.45}$$

where the thrust F and the spacecraft mass m are both assumed to be constant. Each rocket fuel has a specific impulse S_{sp}, which is a (constant) fuel characteristic and defined as

$$S_{sp} = \frac{c}{g_o} = \frac{F}{\dot{m}_f g_o} \tag{4.46}$$

with $g_o = 9.80665$ m/s^2. The fuel m_f consumed can now be obtained from equations (4.46) and (4.45) as

$$m_f = \frac{m}{S_{sp}g_o} \Delta v \tag{4.47}$$

and is seen to be directly proportional to the velocity kick Δv. This explains the great importance of the magnitude of Δv, and the desire to keep Δv as small as possible.

To get a feeling for typical values involved, let us take a geosynchronous satellite ($m = 300$ kg), which is to be de-orbited by two velocity kicks $\Delta v = 3$ m/s each. The onboard thruster uses hydrazine ($S_{sp} = 423$ s in vacuum) as a fuel. The fuel mass required for the de-orbiting manoeuvre is consequently

$$m_f = \frac{300}{423(9.80665)}3 = 0.217 \text{ kg}$$

an amount that, by the way, justifies the assumption $m = $ constant made in the derivation of equation (4.47).

4.9   Energy Increase for Hohmann Transfer

If one assumes that the mass loss of a satellite (due to the fuel consumed by the rocket burst for the impulse) during a velocity kick is negligible, then the potential energy of a satellite remains the same during the velocity kick, since neither the satellite's mass nor its distance from the master change.

During the first velocity kick, the kinetic energy is increased from

$$T_0 = \frac{1}{2}mv_0^2 \tag{4.48}$$

to

$$T_1 = \frac{1}{2}mv_1^2 = mv_0^2 \frac{2}{1+\dfrac{r_1}{r_2}} \tag{4.49}$$

i.e. by the amount

$$\Delta T_1 = T_1 - T_0 = \frac{1}{2}mv_0^2 \frac{1-\dfrac{r_1}{r_2}}{1+\dfrac{r_1}{r_2}} \tag{4.50}$$

Immediately prior to the second velocity kick, the satellite's kinetic energy is

$$T_3 = \frac{1}{2}mv_2^2 = \frac{1}{2}mv_3^2 \frac{2}{1+\dfrac{r_1}{r_2}} \tag{4.51}$$

and immediately thereafter, it is

$$T_3 \; = \; \frac{1}{mv_3^2} \tag{4.52}$$

The increase of kinetic energy during the second velocity kick can thus readily be shown to be

$$\Delta T_2 \; = \; T_3 \; - \; T_2 \; = \; \frac{1}{2}mv_0^2 \, \frac{r_1}{r_2} \, \frac{1 - \dfrac{r_1}{r_2}}{1 + \dfrac{r_1}{r_2}} \tag{4.53}$$

The total kinetic energy increase is

$$\Delta T \; = \; \Delta T_1 \; + \; \Delta T_2 \; = \; \frac{1}{2}mv_0^2\left[1 - \frac{r_1}{r_2}\right] \tag{4.54}$$

This is equal to the total energy which must be imparted to the satellite to move it from a lower to a higher orbit.

$$\Delta E \; = \; \frac{1}{2}mv_0^2\left[1 - \frac{r_1}{r_2}\right] \; = \; \frac{\mu m}{2}\left[\frac{1}{r_1} - \frac{1}{r_2}\right] \tag{4.55}$$

For the first special case, where $r_1 \; = \; r_2$, the energy increase required is obviously zero. For the second special case, where $\Delta v \; = \; \Delta v_{max}$, the energy increase required is

$$\Delta E \; = \; \frac{1}{2}mv_0^2\left[1 - \frac{1}{15.5817}\right] \; = \; 0.468 \, mv_0^2 \tag{4.56}$$

For the third special case, where transfer takes place onto a parabola by a single velocity kick, the energy increase required is

$$\Delta E \; = \; 0.5 \, mv_0^2 \tag{4.57}$$

4.10 Mass Losses Considered

If the mass losses Δm_1 and Δm_2 due to the fuel spent for the impulses to produce the velocity kicks Δv_1 and Δv_2 are considered, and the original mass is designated m_0, the satellite has a mass of $m_0 \; - \; \Delta m_1$ after the first velocity kick, and a mass of $m_0 \; - \; \Delta m_1 \; - \; \Delta m_2$ after the second velocity kick. The increases of kinetic energy during the first velocity kick is then

$$\Delta T_1 \; = \; \frac{1}{2}m_0v_0^2 \, \frac{1 - \dfrac{r_1}{r_2} - \dfrac{2\Delta m_1}{m_0}}{1 + \dfrac{r_1}{r_2}} \tag{4.58}$$

The increase during the second velocity kick is

$$\Delta T_2 \; = \; \frac{1}{2}m_0v_0^2 \, \frac{r_1}{r_2} \cdot \frac{1 - \dfrac{r_1}{r_2} - \dfrac{\Delta m_1}{m_0}\left[1 - \dfrac{r_1}{r_2}\right] - \dfrac{\Delta m_2}{m_0}\left[1 + \dfrac{r_1}{r_2}\right]}{1 + \dfrac{r_1}{r_2}} \tag{4.59}$$

The total kinetic energy increase $\Delta T \; = \; \Delta T_1 \; + \; \Delta T_2$, and is

$$\Delta T = \frac{1}{2} m_0 v_0^2 \left[1 - \frac{r_1}{r_2} - \frac{\dfrac{\Delta m_1}{m_0}\left[2 + \dfrac{r_1}{r_2} - \left(\dfrac{r_1}{r_2}\right)^2 \right] + \dfrac{\Delta m_2}{m_0}\dfrac{r_1}{r_2}\left[1 + \dfrac{r_1}{r_2} \right]}{1 + \dfrac{r_1}{r_2}} \right] \tag{4.60}$$

The mass losses during the velocity kicks cause an increase of potential energy. Just prior to the velocity kick, the potential energy of the satellite is

$$U_0 = -\frac{\mu m_0}{r_1} \tag{4.61}$$

and immediately thereafter, because of the mass loss Δm_1,

$$U_1 = -\frac{\mu(m_0 - \Delta m_1)}{r_1} \tag{4.62}$$

The increase of potential energy during the first velocity kick is then

$$\Delta U_1 = U_1 - U_0 = \frac{\mu \Delta m_1}{r_1} \tag{4.63}$$

In the apoapsis of the transfer orbit, the potential energy is

$$U_2 = -\frac{\mu(m_0 - \Delta m_1)}{r_2} \tag{4.64}$$

and immediately after the second velocity kick, it is

$$U_3 = -\frac{\mu(m_0 - \Delta m_1 - \Delta m_2)}{r_2} \tag{4.65}$$

The increase of potential energy due to the second mass loss during the second velocity kick is thus

$$\Delta U_2 = U_3 - U_2 = \frac{\mu \Delta m_2}{r_2} \tag{4.66}$$

Altogether the potential energy has increased, because of the two mass losses, by

$$\Delta U = \Delta U_1 + \Delta U_2 = \frac{\mu \Delta m_1}{r_1} + \frac{\mu \Delta m_2}{r_2} = \frac{\mu m_0}{r_1}\left[\frac{\Delta m_1}{m_0} + \frac{\Delta m_2}{m_0}\frac{r_1}{r_2} \right] \tag{4.67}$$

With the aid of the velocity equation, one obtains $m_0 v_0^2/2 = \mu m_0/2r_1$. It is then possible to calculate the total energy change for a satellite, displaced from a lower orbit to a higher orbit, by adding the increases in kinetic and potential energy

$$\Delta E = \Delta T + \Delta U = \frac{\mu m_0}{2r_1}\left[1 - \frac{r_1}{r_2}\left(1 - \frac{\Delta m_1 + \Delta m_2}{m_0} \right) \right] \tag{4.68}$$

The same result is, of course, obtained by a direct application of energy equation (2.35)

$$\Delta E = E_3 - E_0 = -\frac{\mu(m_0 - \Delta m_1 - \Delta m_2)}{2r_2} + \frac{\mu m_0}{2r_1} \tag{4.69}$$

as can easily be shown.

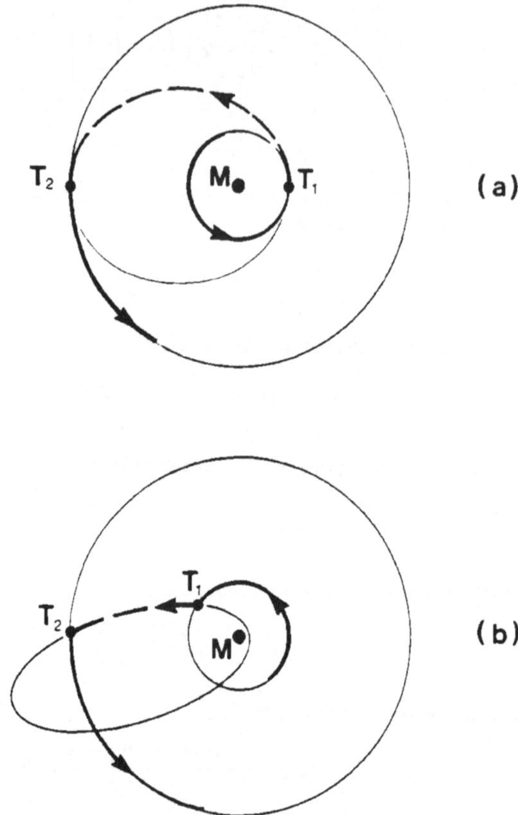

FIGURE 4.7. Hohmann transfer and a transfer with a shorter transfer time.

4.11 Transfer Time for Hohmann Transfer

The duration τ_T of a Hohmann transfer, i.e. the period between departure from the original orbit (point 1 in Figure 4.5) and arrival at the final orbit (point 2 in Figure 4.5), is given by half the Kepler period of the transfer ellipse.

$$\tau_T = \pi\sqrt{\frac{a^3}{\mu}} = \pi\sqrt{\frac{(r_1 + r_2)^3}{8\mu}} \tag{4.70}$$

The advantage of a minimal thrust energy requirement for a Hohmann transfer (Figure 4.7a) is offset by the fact that Hohmann transfers take longer than other types of transfer (Figure 4.7b).

4.12 Several Impulses, Continuous Impulse

Often it is advantageous to sequence *several* individual impulses. For the Hohmann transfer, we have learned that *two* impulses are required. Two impulses also suffice for transfers with shorter transfer times (Figure 4.7). *Several* impulses, following one another, are applied for *course corrections*. The sooner an impulse is applied, the greater is its effect at the target. This is the reason why course corrections are made as early as possible. As soon as a new course error is discovered, another impulse is given. Lawden (1963) has shown, that *optimum* conditions apply, when the time intervals between course corrections become shorter towards the target, and that at a ratio of $e:1$ for subsequent time intervals.

Another possibility to influence the orbit of a satellite is by means of a *continuous* impulse, a process which belongs into the area of *rocket mechanics*, since the mass of the satellite is then changing continuously because of the rocket fuel discharged. If the mass loss is, however, sufficiently small such that a *constant* satellite mass can be assumed, than the continuous impulse may be treated as a small *perturbation* superimposed on the motion in central force field and can be treated, e.g. by the method of the variation of the elements which is presented in a later chapter.

4.13 Launch Windows

A *launch window* is a time interval during which all take-off specifications for a given mission are satisfied.

If a launch window is missed, say due to a malfunction discovered during the countdown, the launch must be postponed until a new launch window opens. The frequency of launch window openings can be determined very readily in case of coplanar circular orbits of two bodies, orbiting the same master, as is approximately true for a launch window on Earth for a mission to intercept, say, Jupiter.

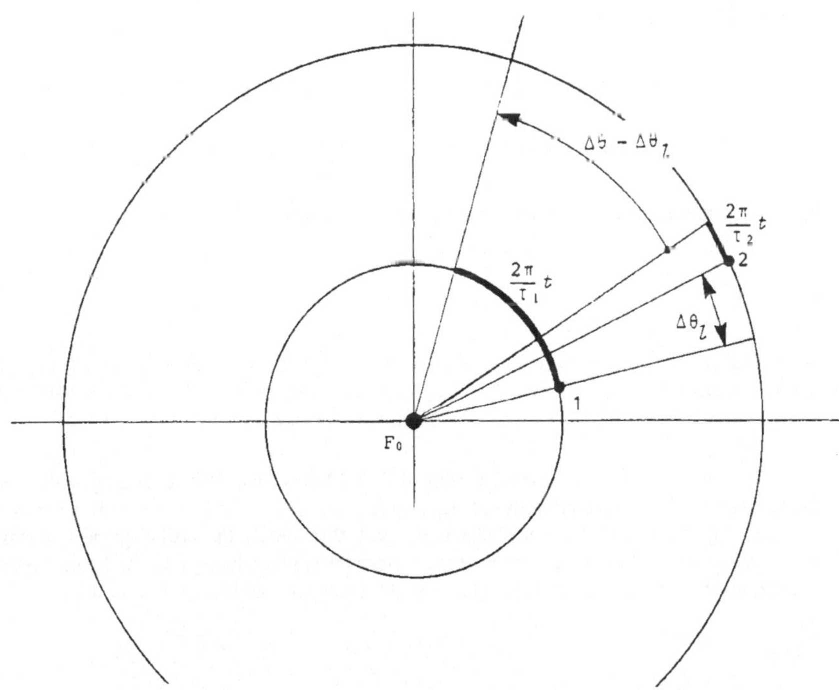

FIGURE 4.8. Two bodies on circular orbits about the same master.

Let τ_1 be the orbital period of body 1 (e.g. Earth), and τ_2 the orbital period of body 2 (e.g. Jupiter) about the common master (e.g. Sun) at F_o, then the angle $\Delta\theta$ (Figure 4.8) of relative position of body 2 with respect to body 1 is

$$\Delta\theta = \frac{2\pi}{\tau_1}t - \frac{2\pi}{\tau_2}t \tag{4.71}$$

If a launch window is open at, say $\Delta\theta_l$, then it will open again when $\Delta\theta$ has reached the value $\Delta\theta_l \pm 2\pi$.

The associated time span is the synodic period τ_{syn}.

$$\tau_{syn} = \frac{\tau_1 \tau_2}{|\tau_2 - \tau_1|} \tag{4.72}$$

The synodic[*] period is the time required for two orbiting bodies to reach the same angular position with respect to each other again.

It is interesting to note, that the frequency of launch window openings for planetary mission is higher for distant planets than for close ones.

The synodic period for Earth-Jupiter missions is e.g.

$$\tau_{syn} = \frac{(1)\ 11.86}{11.86 - 1} = 1.092\ a$$

For Earth-Mars missions, it is

$$\tau_{syn} = \frac{(1)\ 1.881}{1.881 - 1} = 2.135\ a$$

Suggested Reading

1. Hohmann, W. *Die Erreichbarkeit der Himmelskörper*, R. Oldenbourg-Verlag, 1925, 88 pp.

2. Lawden, D.F. *Optimal Trajectories for Space Navigation*, Butterworth, London, 1963, 125 pp.

3. Ting, L. "Optimum Orbital Transfer by Impulses", *ARS Journal*, 30, 11, 1960, pp. 1013-1018.

Problems

4.1   A satellite orbits the Earth, with $a = 8\ 000$ km and $\varepsilon = 0.1$. At a true anomaly of 80° the satellite is given a velocity kick of 500 m/s in a radial direction. Compute (a) the new semi-major axis; (b) the new eccentricity; (c) the new true anomaly; and (d) make a sketch of the old and new orbits.

4.2   A satellite of the Earth has a circular orbit of 7 000 km radius. The ascending node has a right ascension of 100°. The inclination of the orbital plane is 60°. When the orbit angle is 50°, the satellite is given a velocity kick of 0.3 km/s, such that merely the orbital plane is changed, not however the size and the shape of the orbit. Determine (a) the tilt angle κ; (b) the new right ascension Ω; (c) the new inclination i_2; (d) the new orbit angle u_2; and (e) make a sketch.

4.3   Given

$$\sqrt{2} = \sqrt{\frac{2}{1+x}}(1-x) + \sqrt{x}$$

Find x.

4.4   Given

$$y = \sqrt{\frac{2}{1+x}}(1-x) + \sqrt{x} - 1$$

(a) For which value of x has y a maximum? (b) What is y_{max}?

---

[*] synode = gk. "same way" = meeting

4.5   A point satellite, with $a = 14\ 000$ km, $\varepsilon = 0.5$, $m = 200$ kg, orbits the Earth sphere. (a) A tangential velocity kick $\Delta v = 0.4$ km/s is applied in the perigee. Determine the new orbit ($a_2 = ?$, $\varepsilon_2 = ?$) and the increase in energy. (b) If the same velocity kick is applied in the apogee instead of the perigee, what then is the new orbit ($a_2 = ?$, $\varepsilon_2 = ?$) and the energy increase? Make (c) a sketch and compare!

4.6   Given is the Moon ($m = 73.5 \cdot 10^{21}$ kg) on its regular orbit ($a = 384\ 400$ km, $\varepsilon = 0.0549$). You are an engineer in a group making plans to "circularize" the Moon's orbit. A velocity kick is to be applied in the apogee, by means of which the Moon is to assume a circular orbit. Make a sketch and determine (a) the velocity kick required. (b) If the rockets used to impart the velocity kick burn for one minute and provide a constant thrust, what total thrust force is required to place the Moon on a circular orbit? (c) What is the energy increase of the Moon's motion? (d) Considering today's technology, do you deem a circularization of the Moon's orbit feasible? Explain! (Pickering Nuclear Power Station 2160 MW). (e) What is the Moon's Kepler period (on its present orbit)?. (f) What would the Kepler period be, after the circularization? (g) What would the benefits of a circularization of the Moon's orbit be, in your opinion?

4.7   A satellite of the Earth sphere moves in an orbital plane of 5° inclination. Its orbit is circular and has a radius of 6 600 km. The ascending node is at a right ascension of 10°. When the satellite's orbit angle is 45° the satellite is given a velocity kick of 0.4 km/s, such that the orbital plane assumes a new orientation, while the size and the shape of the orbit remain the same. Make (a) a sketch, and determine (b) the tilt angle κ, (c) the new right ascension Ω, (d) the new inclination i_2, and (e) the new orbit angle u_2.

4.8   A satellite of 300 kg mass circles the Earth on a parking orbit, at an altitude of 629 km. It is to be transferred to a coplanar circular orbit of 2 629 km altitude above the Earth sphere's surface. Determine (a) the Hohmann transfer time, (b) the characteristic velocity, (c) the total energy gain, and (d) make a sketch.

4.9   A point satellite orbits a point master on a circular orbit. By comparing the Δv's required, establish whether it is easier (a) to send the satellite off to infinity, or (b) to bring it down.

4.10   By assuming circular orbits about the Sun for the planets, calculate approximate durations for a trip (a) from Earth to Mercury ($r_2 = 57.91 \cdot 10^6$ km); (b) from Earth to Venus ($r_2 = 108.21 \cdot 10^6$ km); (c) from Earth to Mars ($r_2 = 227.9 \cdot 10^6$ km); (d) from Earth to Jupiter ($r_2 = 778.3 \cdot 10^6$ km); (e) from Earth to Saturn ($r_2 = 1.427 \cdot 10^9$ km); (f) from Earth to Uranus ($r_2 = 2.870 \cdot 10^9$ km); (g) from Earth to Neptune ($r_2 = 4.497 \cdot 10^9$ km); and (h) from Earth to Pluto ($r_2 = 5.946 \cdot 10^9$ km). Earth's mean distance from the Sun is $r_1 = 149.6 \cdot 10^6$ km. (The mean distance is equal to the orbit's semi-major axis). Make a sketch for each transfer, and indicate the planet's position at the spacecraft's launch, and the Earth's position at the spacecraft's encounter with the planet.

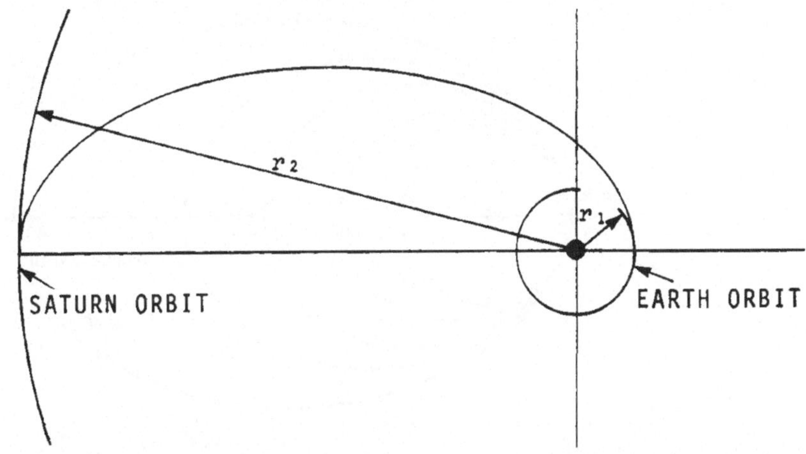

4.11   During the apogee raising phase from an original parking orbit of 7004 km radius to a geosynchronous orbit, a satellite ($m = 2415$ kg) fires its perigee kick motor six times. If the six apogee raises $2\Delta a$ are of equal size, if the satellite mass loss is negligible, and if the thrusters are fired for duration of one minute each time, determine (a) the thrust required for each apogee raise, (b) the eccentricity of each intermediate orbit, and (c) the thrust required by the apogee kick motor which also burns for one minute. For the motors assume that the thrust reaches its constant steady-state force level instantaneously upon the "on" command, and that the tail-off of thrust after the "off" command may be ignored.

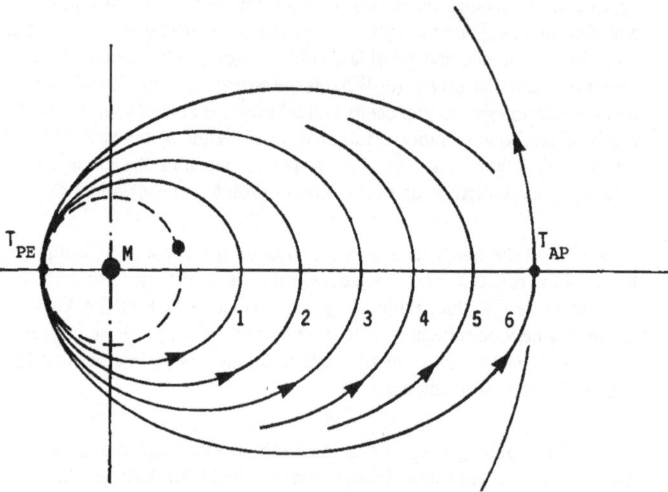

4.12   The sketch shows a proposed six burn apogee raising sequence for a satellite, from a parking orbit of 6771 km radius to the geosynchronous height, i.e. an apogee distance of 42 164 km from Earth centre. (a) Why are the apogee distance raises not equal? (b) What increases would you propose for each of the six raises, and why? (c) What is your explanation for the reasons that favour a six-burn, rather than a single burn, apogee raising sequence? The required thrust is obtained by an impulse of very short duration provided in the perigee of each orbit by the satellite's perigee propulsion system. Assume that the satellite's mass loss at each burn is negligible.

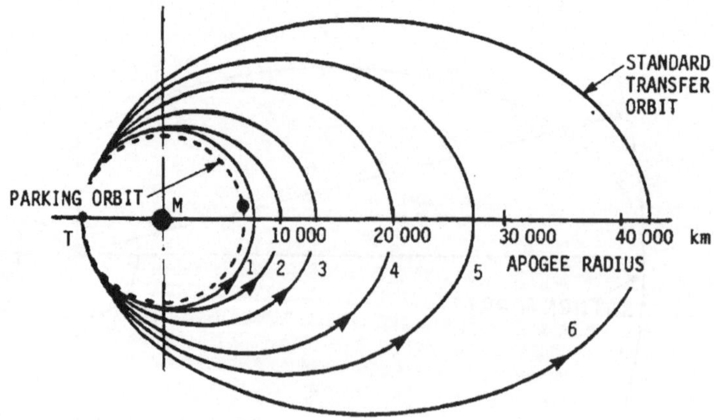

4.13   Anik C-3 ($\uparrow$1982-November-11) plus its PAM (Payload Assist Module) had a combined mass of 3320 kg, which includes 1960 kg of solid propellant in the PAM. Assume a constant spacecraft mass (i.e. structure + 0.5 propellant). (a) What (constant) thrust must the perigee kick motor produce for transfer from a circular parking orbit at 289 km altitude, to an elliptical transfer orbit whose apogee is at geosynchronous altitude, if the perigee kick motor burns for 83 s? Make (b) a sketch to scale. The PAM is separated during coasting on the transfer orbit. Anik C-3, including 490 kg solid propellant for the integrated apogee kick motor, has a mass of 1126 kg. Assume again a constant spacecraft mass (structure + 0.5 propellant), and (c) calculate the (constant) thrust of the apogee kick motor used to circularize the orbit at geosynchronous altitude. Assume that the apogee kick motor also burns for 83 s.

4.14   When inserting a satellite into a geosynchronous orbit, engineers generally feel that it is preferable to make orbital plane corrections at geosynchronous altitude, rather than at low Earth altitudes. To investigate the point, assume a satellite of 3320 kg mass in a circular low Earth parking orbit of 289 km altitude. The orbital plane has an inclination of 10° that must be reduced to 0°. (a) What is the single velocity kick required? (b) What is the (constant) thrust needed if the rocket burn lasts 30 s? Upon arrival at geosynchronous altitude, the satellite has a mass of only 636 kg, including some 150 kg fuel. (c) What is the velocity kick required to effect a 10° orbital plane change? (d) What is the (constant) thrust needed if the rocket burn lasts 30 s, and the satellite mass is assumed to be (a constant) 636 kg? (e) Considering that the fuel (about 150 kg) for the manoeuvre at geosynchronous altitude must be carried to that altitude, which of the two approaches to orbital plane correction is preferable?

4.15   A satellite of 636 kg mass circles the Earth on a parking orbit, at an altitude of 289 km. It is to be transferred to a coplanar orbit at geosynchronous altitude. Determine (a) the Hohmann transfer time, (b) the characteristic velocity, (c) the total energy gain for the 636 kg mass, and (d) make a sketch.

4.16   A satellite ($m = 300$ kg) on a circular orbit ($r = 7000$ km, $i = 30°$) about the Earth is given a sudden velocity kick $\Delta v_z = 0.5$ km/s at the moment when it passes the ascending node on its orbit. Calculate (a) the new transverse velocity, (b) the new orbit inclination angle, (c) the new semi-major axis, and (d) the new eccentricity.

4.17   A spacecraft of 400 kg mass circles the Moon on a parking orbit at an altitude of 20 km. A Hohmann transfer is to take it to a new circular orbit 320 km above the Moon's surface. Determine (a) the transfer time, (b) each of the two velocity kicks, (c) the characteristic velocity, (d) the total energy gain, and (e) make a sketch.

4.18   A geosynchronous satellite is to be de-orbited (i.e. placed into a slightly higher orbit to make room for a new geosynchronous satellite), by three velocity kicks in the direction of motion, 180° apart. The first kick takes place at $\theta = 0°$ with $\Delta v_1 = 2$ m/s, the second kick at $\theta = 180°$ with $\Delta v_2 = 4$ m/s, and third kick at $\theta = 360°$. (a) What Δv_3 is required to ensure that the final orbit is circular? (b) How much higher than geosynchronous is the new orbit? (c) What is the total fuel consumption m_f, if the spacecraft mass ($m = 200$ kg) is assumed to remain constant throughout, and the thruster delivered altogether 250 pulses of 0.5 s duration each during the first velocity kick. The specific impulse S_{sp} of the thruster fuel is 423 s (Thrust = $S_{sp}\dot{m}_f g_o$, with $g_o = 9.80665$ m/s^2).

4.19   The de-orbiting of the geosynchronous Geos-2 ($\uparrow$ 1978, $\uparrow\uparrow$ 1984) in 1984 took place by means of a Hohmann type transfer with three velocity kicks in the direction of motion, 180° apart. If the kicks were $\Delta v_1 = 2.55$ m/s, $\Delta v_2 = 4.93$ m/s, and $\Delta v_3 = 2.43$ m/s, what is (a) the final orbit? (*Hint:* Write $v^2 = \mu\left[\dfrac{2}{r} - \dfrac{1}{a}\right]$ and differentiate with r = constant, to obtain Δa.) (b) What (constant) thrust was required for Δv_1 if the thruster delivered 281 pulses of 0.5 s duration each and the spacecraft mass was 273 kg? Make (c) a sketch, and (d) select that of the three new orbits which is most closely circular.

4.20   A satellite ($m = 1000$ kg) orbiting the Earth on an elliptic path with $r_{PE} = 6570$ km and $r_{AP} = 12\,000$ km, is to be transferred to the geosynchronous orbit ($r_2 = 42\,160$ km) by means of a Hohmann transfer. (i) Calculate (a) the first, (b) the second velocity kick, (c) the characteristic velocity, (d) the transfer time, (e) the (exact) kinetic energy changes at the first, (f) at the second kick, and (g) the total kinetic energy change, all for launch from the perigee of the original orbit. Assume no mass losses. (ii) Repeat the calculations for launch from the apogee of the original orbit. Compare.

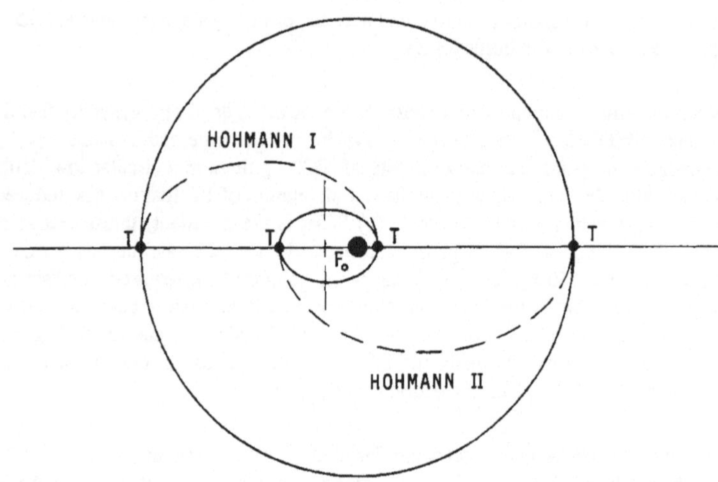

4.21   Calculate the characteristic velocity Δv, (a) for the monoelliptic Hohmann transfer ($1 \rightarrow 2$) from Earth orbit to Neptune orbit ($r = 4\,493 \cdot 10^6$ km), (b) for a bielliptic Hohmann transfer ($1' \rightarrow 3 \rightarrow 2$) from Earth orbit - up to a point at $r_3 = 10\,000 \cdot 10^6$ km - then to Neptune orbit, (c) for a bielliptic (degenerated into biparabolic) Hohmann transfer ($1' \rightarrow \infty \rightarrow 2$) from Earth orbit - via infinity - to Neptune orbit, and (d) calculate the transfer times for the three cases. Assume that Earth orbit and Neptune orbit are both circular and that they are coplanar.

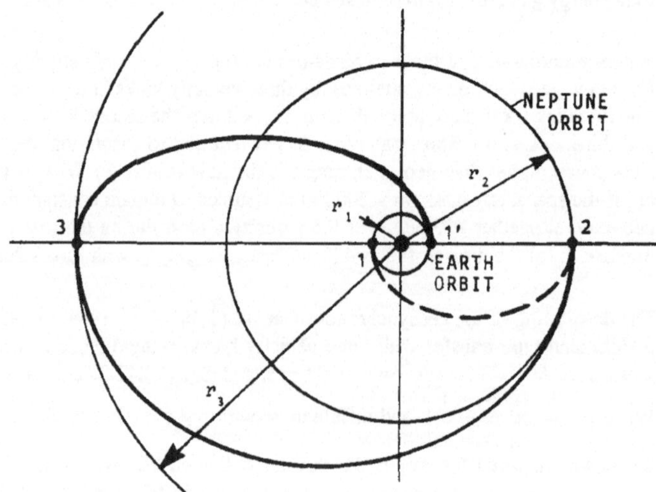

5

The Gravitational Potential

The gravitational attraction force is a *conservative* force, i.e. it can be represented by a *potential*. A *potential* surrounds the *source* mass of the gravitational attraction force. We shall refer to this source as the *master*. If a second mass, which we shall refer to as the *satellite*, is located in the potential of the master, it will be attracted by the master with an attraction force, whose magnitude and direction is defined by the potential at the location of the satellite. The gravitational potential surrounding a *point* master is

$$U = -\frac{\mu m}{r}$$

for a *point* satellite in the gravitational field of the point master. On the following pages, we will establish the gravitational potential surrounding a spheroid master; have a look at a general expression for the potential surrounding a master of arbitrary shape, which is then reduced to the case of a master of rotational symmetry. The results can be found in Table 5.1. Eventually we will introduce the concepts of flattening and perturbation force.

5.1 Approximate Potential Surrounding a Body of Arbitrary Shape

In investigating the behaviour of satellites, we had so far assumed, that the master mass was of *point* size. As we shall show soon, the same results are obtained when the master mass is a *sphere*. A point, or a sphere with spherical mass distribution, represent an extremely good first approximation for a typical master mass, be it Earth, Sun, Moon, or another celestial body. If needed, a second approximation can be provided involving the principal inertia moments of the master. In Figure 5.1 a large body of total mass M is shown, consisting of mass elements dM, located at a radius r' from the origin O of the coordinate system $OXYZ$, which is the principal axes system of the body. The origin O is located at the mass centre of the body. A small mass m of point size is located at a position r from the origin O of the coordinate system, and at a position r'' from the mass element dM. From here on, we shall designate the large mass M as the *master* and the small point mass m as the *satellite*. We thus have $m \ll M$. We also have $|\mathbf{r'}| < |\mathbf{r}|$, as we limit our discussion to satellites located outside the master mass.

Each element dM of the master mass attracts the point satellite m. The elemental potential surrounding each mass element dM, and affecting the point mass m, is

$$dU = -\frac{G\,dM\,m}{r''} \tag{5.1}$$

For the whole master mass M, the potential becomes

$$U = -G\,m\int_M \frac{dM}{r''} \tag{5.2}$$

Let the angle between the position vectors r and r' (Figure 5.1) be ϕ. According to the cosine law, we thus have

$$r''^2 = r^2 + r'^2 - 2rr'\cos\phi \tag{5.3}$$

Which may be rewritten in the form

$$\frac{1}{r''} = \frac{1}{r}\frac{1}{\sqrt{1 + \frac{r'^2}{r^2} - 2\frac{r'}{r}\cos\phi}} \tag{5.4}$$

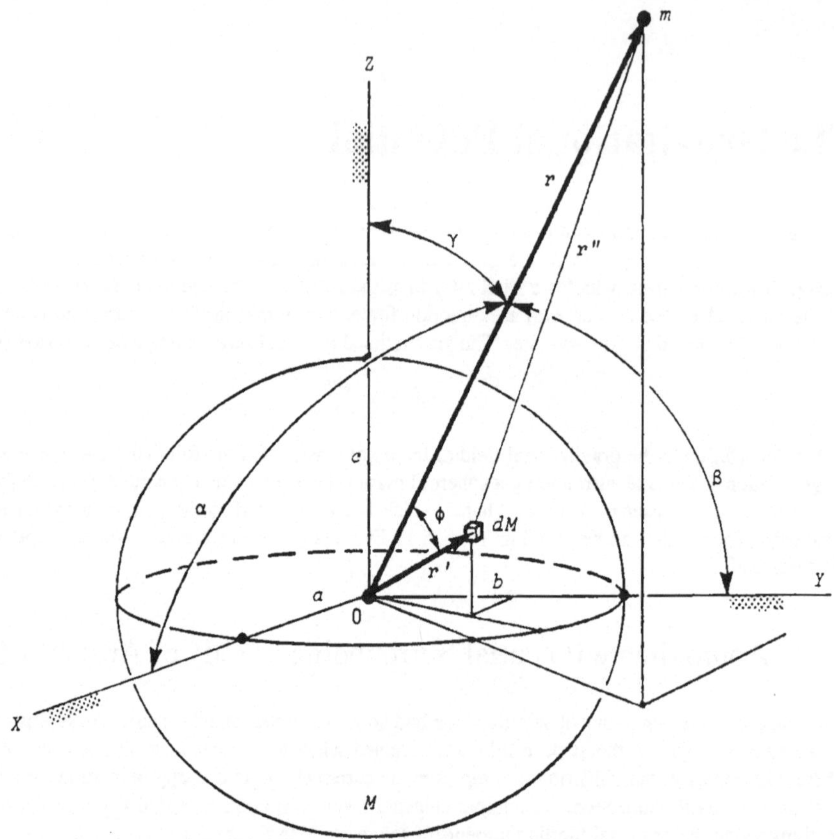

FIGURE 5.1. Point satellite m and ellipsoidal master M.

Since $r' < r$ for all satellites, we can expand into a series for (r'/r)

$$\frac{1}{r''} = \frac{1}{r}\left\{ 1 + \frac{r'}{r}\cos\phi + \left[\frac{r'}{r}\right]^2 \left[\frac{3}{2}\cos^2\phi - \frac{1}{2}\right] + \cdots \right\} \tag{5.5}$$

As will be shown later, a very good approximation suitable even for satellites orbiting close to their master, is obtained by breaking off the series after the term with $(r'/r)^2$.

We write

$$r^2 = X^2 + Y^2 + Z^2 \tag{5.6}$$

and

$$\cos\phi = \frac{\mathbf{r}\cdot\mathbf{r}'}{rr'} = \frac{XX' + YY' + ZZ'}{rr'}$$

which, after introduction of the direction cosines

$$\cos\alpha = \frac{X}{r} \qquad \cos\beta = \frac{Y}{r} \qquad \cos\gamma = \frac{Z}{r}$$

becomes

$$\cos\phi = \frac{1}{r'}(X'\cos\alpha + Y'\cos\beta + Z'\cos\gamma) \tag{5.7}$$

We further conclude that all inertia products, of the type $\int X'Y'dM$, will vanish, since $OXYZ$ represents a system of principal axes of the master mass M, and can then write for the potential

$$U = -\frac{Gm}{r} \int_M \left\{ 1 - \frac{1}{2}\frac{1}{r^2}(X'^2 + Y'^2 + Z'^2) + \frac{3}{2r^2}\left[X'^2\cos^2\alpha + Y'^2\cos^2\beta + Z'^2\cos^2\gamma\right] \right\} dM$$

$$(5.8)$$

Upon integration and observation that

$$\int_M (X'^2 + Y'^2 + Z'^2)dM = \frac{1}{2}(A + B + C)$$

where A, B and C are the inertia moments of the master about the principal axes, the potential becomes

$$U = -\frac{GMm}{r}\left\{ 1 + \frac{A(1 - 3\cos^2\alpha) + B(1 - 3\cos^2\beta) + C(1 - 3\cos^2\gamma)}{2Mr^2} \right\} \qquad (5.9)$$

One may use a more compact form, viz.

$$U = -\frac{GMm}{r}\left\{ 1 + \frac{A + B + C - 3I_C}{2Mr^2} \right\} \qquad (5.10)$$

where

$$I_C = A\cos^2\alpha + B\cos^2\beta + C\cos^2\gamma \qquad (5.11)$$

is the inertia moment of the master about the line $0m$ between master mass centre and satellite. Equation (5.10) is, of course, only valid *outside* the master.

From equation (5.10), it can be concluded immediately, that the potential field surrounding a master mass of *point* size, is given by

$$U = -\frac{GMm}{r} \qquad (5.12)$$

since the inertia moments vanish for a point, i.e. since $C = B = A = 0$.

The same result is obtained for the potential surrounding a master mass of *spherical* mass distribution, since for a sphere $C = B = A$.

For an axisymmetric master mass, with Z as the axis of symmetry,

$$B = A \qquad (5.13)$$

From Figure 5.2 one obtains

$$\cos\alpha = \cos\delta \cos\lambda \qquad (5.14a)$$

$$\cos\beta = \cos\delta \sin\lambda \qquad (5.14b)$$

$$\cos\gamma = \sin\delta \qquad (5.14c)$$

We also introduce the radius R_e of the equator of the master mass and obtain eventually for the potential around a master mass of axisymmetric shape

$$U = -\frac{GMm}{r}\left\{ 1 - \frac{1}{2}\frac{R_e^2}{r^2}\frac{C - A}{MR_e^2}(3\sin^2\delta - 1) \right\} \qquad (5.15)$$

Since the inertia moments C and A of the Earth (or other celestial bodies) cannot be computed directly, the quantity $(C - A)/MR_e^2$ is often combined into a constant, which is generally referred to as J_2 in celestial mechanics. We also introduce $GM = \mu$, and write now for the potential

$$U = -\frac{\mu m}{r}\left\{ 1 - \frac{1}{2}\frac{R_e^2}{r^2}J_2(3\sin^2\delta - 1) \right\} \qquad (5.16)$$

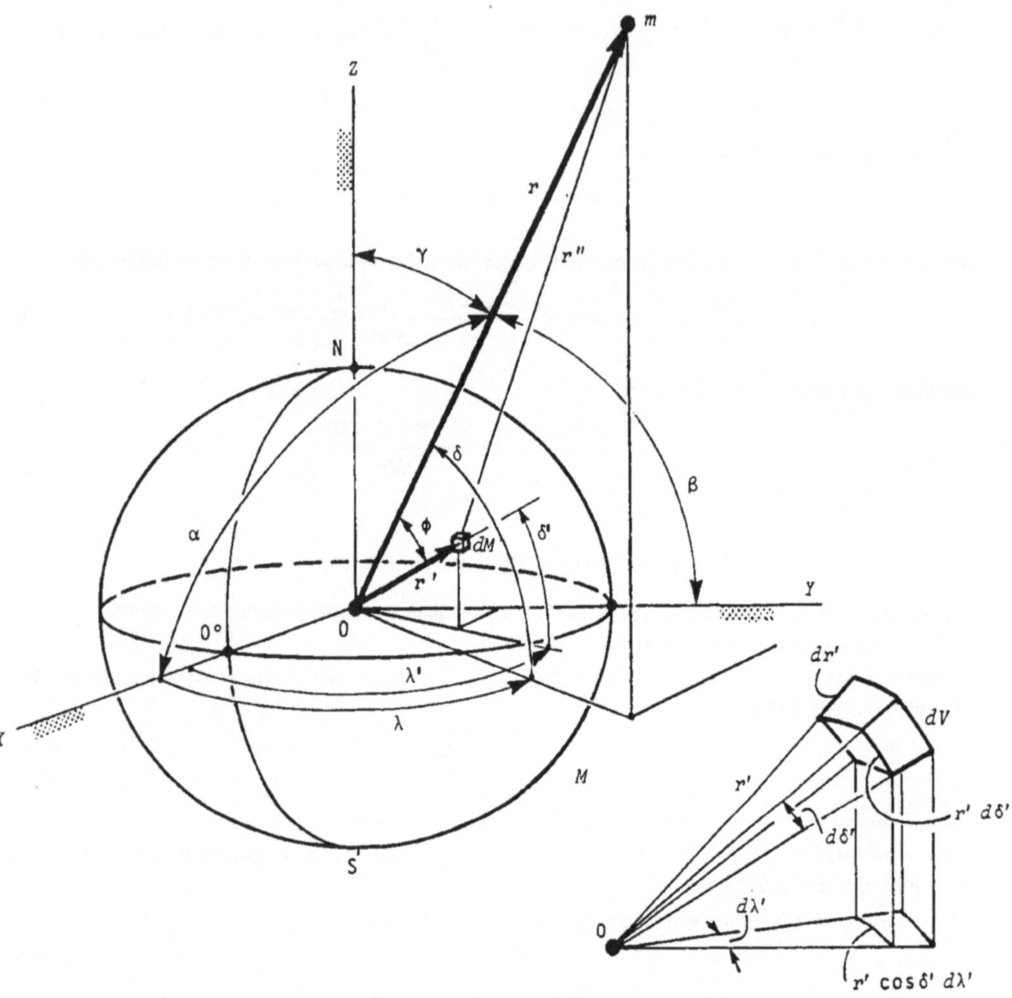

FIGURE 5.2. Coordinates.

Numerical values for the Earth spheroid are

$$R_e = 6378.160 \text{ km}$$

$$J_2 = 1082.628 \cdot 10^{-6}$$

$$\mu = 398\,601.19 \text{ km}^3/\text{s}^2$$

5.2   Potential Surrounding a Body of Arbitrary Shape

In Figure 5.2 a body of arbitrary shape and having a total mass M is shown. It consists of mass elements dM, each of which is located at a distance r' from the origin of the coordinate system, which coincides with the mass centre of the body. A point mass m is located outside of the master mass M at a distance r from the origin of the coordinate system and at a distance r'' from the arbitrary mass element dM. Each master mass element dM attracts the satellite m. The elemental potential surrounding each mass element dM is

$$dU = -\frac{G\,dM\,m}{r''} \tag{5.17}$$

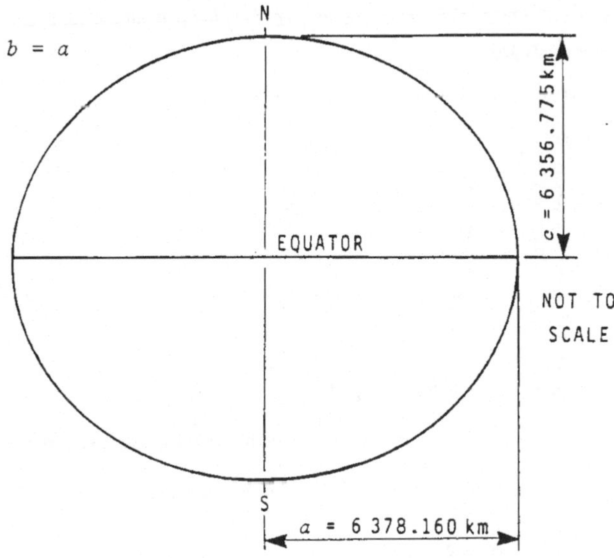

FIGURE 5.3. The rotationally symmetric Earth ellipsoid (= spheroid)

The potential U for the whole master mass M is obtained by integration over the whole master mass

$$U = -Gm \int_M \frac{dM}{r''} \tag{5.18}$$

If we now introduce the density ρ, whose distribution throughout the mass is unknown, and an elemental volume dV, then we obtain for the potential

$$U = -Gm \int_V \frac{\rho dV}{r''} \tag{5.19}$$

The angle between position vector $\mathbf{r}$ pointing to the satellite and position vector $\mathbf{r}'$ pointing to the mass element dM is ϕ (Figure 5.2). Using the cosine law, one gets

$$\frac{1}{r''} = \frac{1}{r} \frac{1}{\sqrt{1 + \frac{r'^2}{r^2} - 2\frac{r'}{r}\cos\phi}} \tag{5.20}$$

We expand equation (5.20) into a series by using Legendre[*] polynomials $P_n(\cos\phi)$.

$$\frac{1}{r''} = \frac{1}{r} \sum_{n=0}^{\infty} \left[\frac{r'}{r}\right]^n P_n(\cos\phi) \tag{5.21}$$

Series (5.21) converges, provided that

$$\frac{r'}{r} \leq 1 \tag{5.22}$$

---

[*] Adrien Marie Legendre (1752-1833), French mathematician.

a condition always satisfied for satellites, since their distance from the centre of the master mass is always greater than the distance of a mass element of the master mass itself from the same centre.

The Legendre polynomials are

$$P_0(\chi) = 1$$
$$P_1(\chi) = \chi$$
$$P_2(\chi) = \frac{3}{2}\chi^2 - \frac{1}{2}$$
$$P_3(\chi) = \frac{5}{2}\chi^3 - \frac{3}{2}\chi \tag{5.23}$$
$$P_4(\chi) = \frac{35}{8}\chi^4 - \frac{30}{8}\chi^2 + \frac{3}{8}$$
$$P_5(\chi) = \frac{63}{8}\chi^5 - \frac{70}{8}\chi^3 + \frac{15}{8}\chi$$

The listing of Legendre polynomials can be continued indefinitely, by employing the recursion formula

$$P_n(\chi) = \frac{2n-1}{n}\chi P_{n-1}(\chi) - \frac{n-1}{n}P_{n-2}(\chi), \qquad n \geq 1 \tag{5.24a}$$

or by means of the Rodrigues[*] formula

$$P_n(\chi) = \frac{1}{2^n n!}\frac{d^n}{d\chi^n}(\chi^2 - 1)^n, \qquad n \geq 1 \tag{5.24b}$$

By using Legendre polynomials (5.24), the potential surrounding the master mass can now be written

$$U = -\frac{Gm}{r}\sum_{n=0}^{\infty}\int_V \rho\left[\frac{r'}{r}\right]^n P_n(\cos\phi)\,dV \tag{5.25}$$

Attractor Mass	Potential (outside of attractor mass)
Point	$U = -\dfrac{\mu m}{r}$
Sphere	$U = -\dfrac{\mu m}{r}$
Body of arbitrary shape (approximation)	$U = -\dfrac{\mu m}{r}\left\{1 + \dfrac{A+B+C-3I_C}{2Mr^2}\right\}$
Spheroid of slight ellipticity	$U = -\dfrac{\mu m}{r}\left\{1 - \dfrac{1}{2}\dfrac{R_e^2}{r^2}\cdot J_2 \cdot (3\sin^2\delta - 1)\right\}$
Body of revolution	$U = -\dfrac{\mu m}{r}\left\{1 - \sum_{n=2}^{\infty}\left[\dfrac{R_e}{r}\right]^n J_n P_n(\sin\delta)\right\}$
Body of arbitrary shape	$U = -\dfrac{\mu m}{r}\left\{1 + \dfrac{1}{M}\sum_{n=2}^{\infty}\int_M\left[\dfrac{r'}{r}\right]^n P_n(\cos\phi)\,dM\right\}$

TABLE 5.1. Potential energy expressions.

---

[*] Olide Rodrigues (1794–1851), French mathematician and economist.

5.3   Potential Surrounding a Body of Revolution

We have established that the potential about a body of arbitrary shape can be represented by an equation (5.25) involving a series expansion. Let us now restrict ourselves to bodies of revolution.

Upon introducing the longitude λ' and the latitude δ' as coordinates (Figure 5.2), one obtains for the volume element dV the expression

$$dV = r'^2 \cos\delta' \, d\delta' \, d\lambda' \, dr' \tag{5.26}$$

Using the dot product of the position vectors of satellite m and the volume element dV of the master, the cosine of the angle ϕ between the two vectors is

$$\cos\phi = \frac{\mathbf{r} \cdot \mathbf{r'}}{rr'} = \frac{xx' + yy' + zz'}{rr'} \tag{5.27}$$

From Figure 5.2, it is concluded that

$$
\begin{aligned}
x &= r \cos \delta \cos \lambda \\
x' &= r' \cos \delta' \cos \lambda' \\
y &= r \cos \delta \sin \lambda \\
y' &= r' \cos \delta' \sin \lambda' \\
z &= r \sin \delta' \\
z' &= r' \sin \delta'
\end{aligned}
\tag{5.28}
$$

Using expressions (5.27) and (5.28), one obtains for the cosine of the angle between the position vectors of satellite and volume element

$$\cos \phi = \sin \delta \sin\delta' + \cos \delta \cos \delta' \cos(\lambda - \lambda') \tag{5.29}$$

Writing each term of the series for the potential (5.25) in full,

$$U = -\frac{Gm}{r}\left[\int_V \rho \, dV + \int_V \rho \, \frac{r'}{r}\cos\phi \, dV + \int_V \rho\left[\frac{r'}{r}\right]^2 \left[\frac{3}{2}\cos^2\phi - \frac{1}{2}\right] dV + \cdots \right] \tag{5.30}$$

Observing that $dM = \rho \, dV$, the first integral in equation (5.30) represents the total mass of the master

$$\int_V \rho \, dV = \int_M dM = M \tag{5.31}$$

The second integral in equation (5.30) can be expressed in the following form, if use is made of equation (5.27),

$$\frac{1}{r^2}\left[x \int_M x' dM + y \int_M y' dM + z \int_M z' dM \right] \tag{5.32}$$

If the mass centre of the master mass M and the origin of the coordinate system coincide, we have

$$\int_M x' dM = \int_M y' dM = \int_M z' dM = 0$$

and the whole term (5.32) vanishes.

By using equation (5.26), we can write for the third integral in equation (5.30)

$$\frac{1}{r^2} \iiint_V \rho r'^4 \left[\frac{3}{2}\cos^2\phi - \frac{1}{2}\right] \cos\delta' \, dr' \, d\lambda' \, d\delta' \tag{5.33}$$

The expression in square brackets can be rewritten in terms of longitude and latitude by employing equation (5.29) and gives

$$\frac{3}{2}\cos^2\phi \; - \; \frac{1}{2} \; = \; \frac{3}{2}\left[\sin^2\delta \sin^2\delta' \; + \; 2\sin\delta \sin\delta' \cos\delta \cos\delta'\cos(\lambda - \lambda')\right.$$

$$\left. + \; \cos^2\delta \cos^2\delta' \cos^2(\lambda - \lambda')\right] \; - \; \frac{1}{2}$$

$$= \; \left[\frac{3}{2}\sin^2\delta - \frac{1}{2}\right]\left[\frac{3}{2}\sin^2\delta' - \frac{1}{2}\right] \qquad\qquad (5.34)$$

$$+ \; 3\sin\delta \sin\delta' \cos\delta \cos\delta'\cos(\lambda - \lambda')$$

$$+ \; \frac{1}{2}\cos^2\delta \cos^2\delta' \cos2(\lambda - \lambda')$$

which, in turn, can be reformed into terms with Legendre polynomials

$$\frac{3}{2}\cos^2\phi \; - \; \frac{1}{2} \; = \; P_2(\sin\delta) \; \cdot \; P_2(\sin\delta')$$

$$+ \; 3\sin\delta \sin\delta' \cos\delta \cos\delta' \cos(\lambda - \lambda') \qquad\qquad (5.35)$$

$$+ \; \frac{1}{2}\cos^2\delta \cos^2\delta' \cos(\lambda - \lambda')$$

Before continuing, let the study be restricted to a master mass with rotational symmetry about the z-axis (Figure 5.3). In that case, all terms containing $\cos(\lambda - \lambda')$ and $\cos2(\lambda - \lambda')$ vanish when the integral with the limits $\lambda' = 0$ and $\lambda' = 2\pi$ is formed. For the third integral of equation (5.30), we thus find

$$\frac{2\pi}{r^2}P_2(\sin\delta) \iint_V \rho r'^4 \cdot P_2(\sin\delta') \cdot \cos\delta' \, dr' \, d\delta' \qquad\qquad (5.36)$$

As long as a detailed distribution of the volume (mountains, valleys) and detailed distribution of the density (ore bodies, oceans, liquid core) are not known, the integral in equation (5.36) cannot be solved. Let us therefore introduce a *coefficient J_2 of the zonal spherical harmonic* defined by

$$J_2 \; = \; -\frac{2\pi}{M R_e^2} \iint \rho r'^4 \, \cos\delta' \, P_2(\sin\delta') \, dr' \, d\delta' \qquad\qquad (5.37)$$

where R_e is the equator radius of the master mass.

In a similar fashion, the remaining terms of equation (5.30) for the potential can be handled and it can be shown that the potential can be expressed by a series

$$U \; = \; \frac{\mu m}{r} \sum_{n=0}^{\infty}\left[\frac{R_e}{r}\right]^n J_n \; \cdot \; P_n(\sin\delta) \qquad\qquad (5.38)$$

The terms J_n are the coefficients of the nth zonal spherical harmonic, where the adjective *zonal* indicates that there is *no* dependence on the longitude λ', i.e. rotational symmetry is pre-supposed. Because of the assumptions made, the coefficient of the zeroth harmonic becomes -1, and the coefficient of the first harmonic vanishes. The coefficients of the higher harmonics for the *geopotential*, i.e. the Earth's gravitational potential, have been determined experimentally, by observing satellite behaviour:

$$J_0 \; = \; -1$$

$$J_1 \; = \; 0$$

$$J_2 \; = \; 1082.628 \; \cdot \; 10^{-6}$$

$$J_3 \; = \; -2.538 \; \cdot \; 10^{-6}$$

etc., see Table 5.2.

J_0	=	-1		J_{11}	=	$0.202 \cdot 10^{-6}$
J_1	=	0		J_{12}	=	$-0.042 \cdot 10^{-6}$
J_2	=	$1082.628 \cdot 10^{-6}$		J_{13}	=	$-0.123 \cdot 10^{-6}$
J_3	=	$-2.538 \cdot 10^{-6}$		J_{14}	=	$-0.073 \cdot 10^{-6}$
J_4	=	$-1.593 \cdot 10^{-6}$		J_{15}	=	$-0.174 \cdot 10^{-6}$
J_5	=	$-0.230 \cdot 10^{-6}$		J_{16}	=	$0.187 \cdot 10^{-6}$
J_6	=	$0.502 \cdot 10^{-6}$		J_{17}	=	$0.085 \cdot 10^{-6}$
J_7	=	$-0.361 \cdot 10^{-6}$		J_{18}	=	$-0.231 \cdot 10^{-6}$
J_8	=	$-0.118 \cdot 10^{-6}$		J_{19}	=	$-0.216 \cdot 10^{-6}$
J_9	=	$-0.100 \cdot 10^{-6}$		J_{20}	=	$-0.005 \cdot 10^{-6}$
J_{10}	=	$-0.354 \cdot 10^{-6}$		J_{21}	=	$0.145 \cdot 10^{-6}$

TABLE 5.2. Coefficients of zonal spherical harmonics of the geopotential (Y. Kozai, 1970).

The coefficients with even numbered subscripts represent a measure for symmetric deviations (with the equatorial plane as plane of symmetry) from a spherical mass, those with odd numbered subscripts give a measure of non-symmetric deviations (e.g. pear shape of the Earth mass distribution).

The assumption of rotational symmetry represents the actual mass distribution of the Earth (and any other large celestial body) very closely. By interpreting satellite behaviour, physicists have computed an ellipticity of the equator represented by a major axis of somewhere around 100 m greater length than the minor axis. The major axis stretches from the mid-Atlantic ($\lambda = -25°$) to the eastern Pacific ($\lambda = 155°$).

The shape of the Earth computed from satellite behavioural data represents the *mass* distribution of the Earth, i.e. the distribution of the product of density and volume. The *geometric* shape of the surface of the Earth may, but need not, be the same.

Measurements and computations for the Earth have shown, that the volume distribution (geometric shape) and the mass distribution (dynamic shape) are in good agreement. The Earth's shape is very close to that of a body in *hydrostatic equilibrium*, i.e. the Earth exhibits approximately the shape of a fluid subjected to its own gravitational and centrifugal forces (see e.g. Bucerius and Schneider, 1967).

5.4 Perturbation Force in Gravitational Potential

The potential surrounding the master can be resolved into two portions, the *Kepler* potential, associated with a *Kepler* force, and a *perturbation* potential, associated with a *perturbation* force.

For the potential (5.15)

$$U = -\frac{\mu m}{r} \left[1 - \frac{R_e^2 J_2}{r^2} P_2 (\sin\delta) \right] \tag{5.39}$$

of a body of revolution, the perturbation portion is

$$U_{pert} = \frac{\mu m R_e^2 J_2}{r^3} P_2 (\sin\delta) \tag{5.40}$$

or, since $P_2(\sin\delta) = \frac{3}{2}\sin^2\delta - \frac{1}{2}$,

$$U_{pert} = \frac{3}{2} \frac{\mu m R_e^2 J_2}{r^3} \left[\sin^2\delta - \frac{1}{3} \right] \tag{5.41}$$

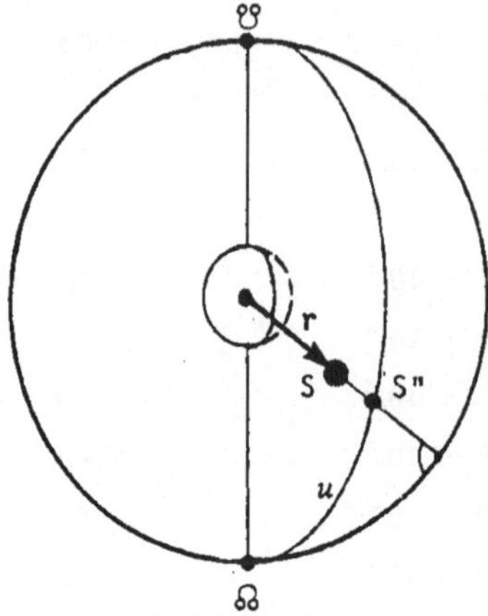

$$u \;=\; \theta \;+\; \omega$$

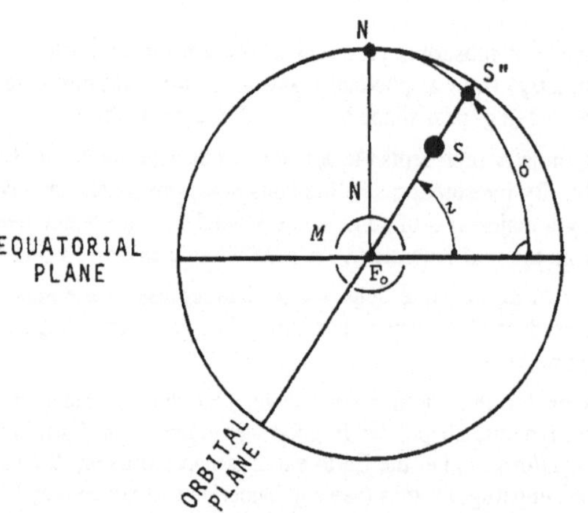

FIGURE 5.4. Celestial sphere.

The perturbation force $\mathbf{f}$ is

$$\mathbf{f} \;=\; -\mathrm{grad}\, U_{pert} \tag{5.42}$$

Let us make use of cylindrical coordinates r, θ, z. In order to obtain the latitude δ of the satellite position, we inspect Figure 5.4, which shows satellite position vector $\mathbf{r}$, satellite orbit angle u, inclination i, and latitude δ, and Figure 5.5, in which equatorial and orbital plane are shown.

The sine law gives

$$\sin\delta \;=\; \sin i \, \sin u \tag{5.43}$$

The elements of the cylindrical coordinates are (Figure 5.5)

$$dr, \quad rd\theta, \quad dz \tag{5.44}$$

with $dz = r \sin u \, di$.

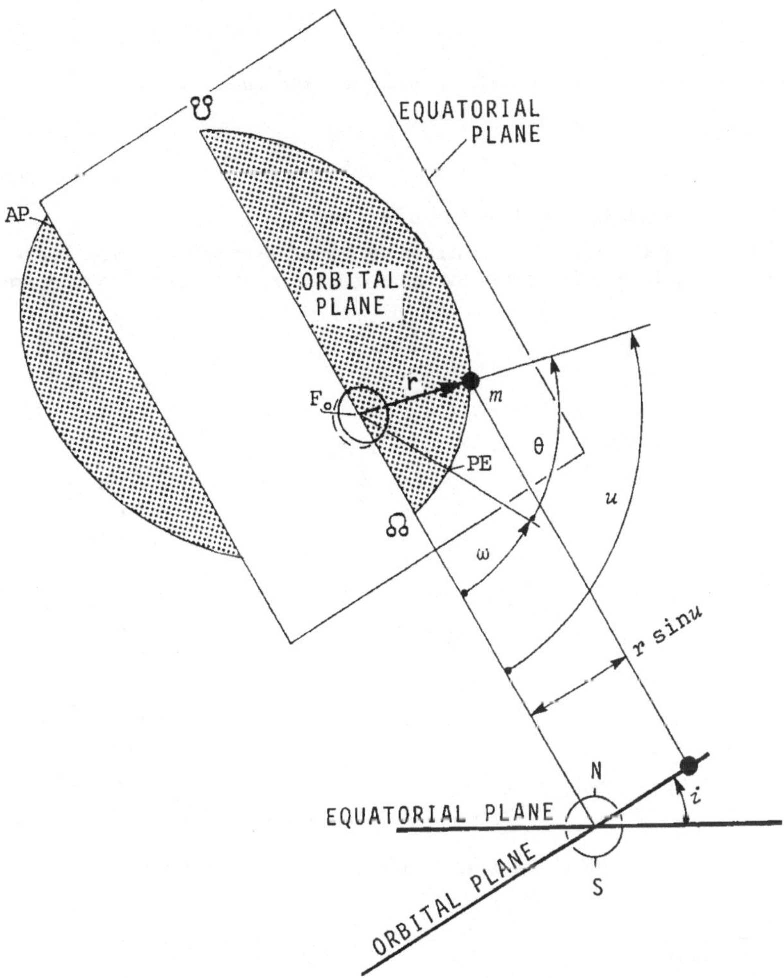

FIGURE 5.5. Equatorial plane and orbital plane.

such that

$$\mathbf{grad} \;=\; \left[\mathbf{e}_r\, \mathbf{e}_\theta\, \mathbf{e}_z\right] \begin{bmatrix} \dfrac{\partial}{\partial r} \\[2ex] \dfrac{\partial}{r\partial\theta} \\[2ex] \dfrac{\partial}{r\,\sin u\,\partial i} \end{bmatrix} \tag{5.45}$$

From equations (5.42) and (5.43), we obtain for the perturbation force

$$\mathbf{f} \;=\; -\mathbf{grad}\left[\frac{3}{2}\,\frac{\mu m R_e^2 J_2}{r^3}\left(\sin^2 i\,\sin^2 u - \frac{1}{3}\right)\right] \tag{5.46}$$

The gradient can be determined with the aid of equation (5.45). The individual components (Figure 5.6) of the perturbation force are then

$$f_r \;=\; -\frac{3\mu m R_e^2 J_2}{2r^4}\left(1 - 3\sin^2 i\,\sin^2 u\right) \tag{5.47}$$

$$f_\theta = -\frac{3\mu m R_e^2 J_2}{2r^4}\sin^2 i\ \sin 2u \tag{5.48}$$

$$f_z = -\frac{3\mu m R_e^2 J_2}{2r^4}\sin 2i\ \sin u \tag{5.49}$$

where $r = p/(1+\varepsilon\cos\theta)$, $u = \theta + \omega$, and $\partial u = \partial\theta$.

The further away a satellite is from its master, the smaller the perturbation force. The perturbation force declines very rapidly as a function of the distance, since the latter appears to the fourth power in the denominator.

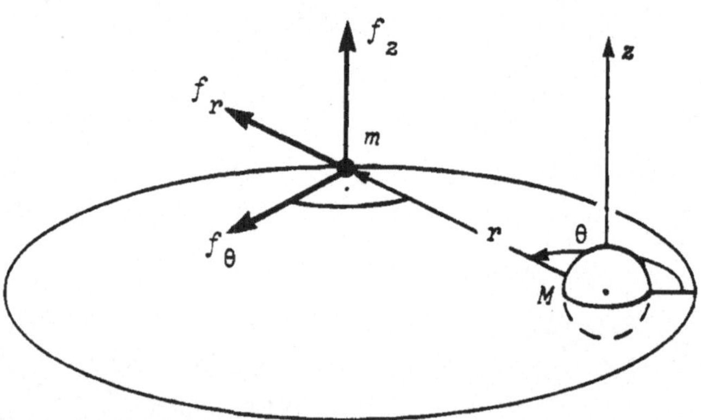

FIGURE 5.6. The components of the perturbation force.

5.5 Flattening

The Earth is *not* a perfect sphere. It exhibits numerous deviations from a perfectly spherical shape. Similar to a *mandarine*, it is *flattened* at the poles. There are other deviations, e.g. negative *bulges* in the northern hemisphere and positive bulges in the southern hemisphere, resembling the shape of a *pear*, with the stem at the north pole.

The *flattening* is however the most important deviation. As mathematical model for the resulting shape a *spheroid* ($=$ ellipsoid of revolution) is used. The equator of the spheroid is circular and so are all sections parallel to the equator. The meridians are ellipses, whose semi-major axes are equal to the equator radius R_e. The semi-minor axes c are the distance from the equator plane to either pole. The *flattening f* is the ratio of the difference $a - c$ of the semi-axes to the equatorial radius a,

$$f = \frac{a-c}{a} = 1 - \frac{c}{a} \tag{5.50}$$

For the Earth the flattening is

$$f = \frac{1}{298.25} \tag{5.51}$$

With an equatorial radius of

$$R_e = a = b = 6378.160\ \text{km} \tag{5.52}$$

the semi-minor axis to the poles becomes

$$R_P = c = a(1-f) = 6356.775\ \text{km} \tag{5.53}$$

The *standard* radius R for the Earth *sphere* is a theoretical measure, obtained by equating the volume $\frac{4}{3}\pi R^2$ of a sphere to the volume $\frac{4}{3}\pi abc$ of an ellipsoid, and then rounding off to the nearest kilometer,

$$R = 3\sqrt{R_e^2 R_p} = 6371 \text{ km} \tag{5.54}$$

The planets Mars, Jupiter and Saturn have flattenings, which are noticeably *greater* than that of the Earth. Particularly large are those of Jupiter and Saturn.

$$\text{Earth } f = 1/298.25 \tag{5.55a}$$

$$\text{Mars } f = 1/199 \tag{5.55b}$$

$$\text{Jupiter } f = 1/15.2 \tag{5.55c}$$

$$\text{Saturn (with rings) } f = 1/10.2 \tag{5.55d}$$

The flattening of the Moon is very small, around $f = 1/3000$. Since an assumption of hydrostatic equilibrium for the (solid) Moon cannot be justified, one would have to differentiate between *geometric* (i.e. shape) and *dynamic* (i.e. mass distribution) flattening. A definition of the flattening of the Moon is further complicated by the fact that the ellipticity of the Moon equator is of the same order of magnitude as the ellipticity of the meridians. Thus it is justifiable engineering practice to consider the Moon's geometric shape as spherical. The Moon's mass distribution is such that there is a slight concentration along an axis pointing to the Earth, as can be seen from the Moon's principal inertia moments. This concentration makes it possible for the Moon to orbit the Earth in *locked rotation*.

5.6   Inertia Moments

Since the Earth has a density which increases towards the centre, a direct calculation of the inertia moments is exceedingly difficult. From satellite behaviour, physicists have computed a (not very exact) value for the inertia moment, which is usually presented in the form (Garland, 1979)

$$\frac{C}{MR_e^2} = 0.330\,78 \tag{5.56}$$

where C is the *inertia moment about the spin axis*, M is the *total mass*, and R_e is the *equator radius* of the Earth. For the inertia moment about the spin axis, we thus obtain

$$C = 0.330\,78\, MR_e^2 \tag{5.57}$$

The spin axis represents one principal axis of the Earth spheroid. Because of the rotational symmetry, the other two principal inertia moments must be equal, $B = A$. Using equations (5.15) and (5.16), we have

$$\frac{C - A}{MR_e^2} = J_2$$

and can thus obtain for the inertia moment

$$A = (0.33078 - J_2)MR_e^2 \tag{5.58}$$

If we use for the Earth's mass

$$M = 5.976 \cdot 10^{24} \text{ kg}$$

for the Earth's equator radius

$$R_e = 6378.16 \text{ km}$$

and for the Earth's second zonal harmonic coefficient

$$J_2 = 1082.628 \cdot 10^{-6}$$

then we obtain

$$MR_{\bullet}^2 \;=\; 2.431 \cdot 10^{32} \text{ km}^2 \text{ kg}$$

and consequently for the Earth's principal inertia moments

$$A \;=\; 80.152 \cdot 10^{30} \text{ km}^2\text{kg} \tag{5.59a}$$

$$B \;=\; A \tag{5.59b}$$

$$C \;=\; 80.416 \cdot 10^{30} \text{ km}^2\text{kg} \tag{5.59c}$$

Inertia moments are readily identified and measured for rigid bodies only. The Earth's interior is known to be liquid, there are huge water masses on the Earth's surface and there is an atmosphere surrounding the Earth. Fortunately, these fluids do follow the motion of the solid Earth crust relatively faithfully. This circumstance makes a determination of inertia moments for the Earth, as given above, possible, but they are only approximate.

The Moon, on the other hand, is a celestial body known to be solid throughout. Its principal inertia moments are thus readily identifiable and amount to

$$A \;=\; 88.027 \cdot 10^{27} \text{ km}^2\text{kg} \tag{5.60a}$$

$$B \;=\; 88.045 \cdot 10^{27} \text{ km}^2\text{kg} \tag{5.60b}$$

$$C \;=\; 88.082 \cdot 10^{27} \text{ km}^2\text{kg} \tag{5.60c}$$

The reference axes are located such that C gives the inertia moment about the Moon's spin axis, A about an axis point towards the Earth, and B about an axis perpendicular to both other axes.

5.7   The Earth

The actual Earth shape is referred to as *geoid*. As approximations for the geoid simple mathematical models are used. The first approximation, and a very good one for numerous applications, is that of a *sphere*. The flattening at the poles is considered when the geoid is represented by a *spheroid* (= ellipsoid of revolution).

A more general representation is obtained by superposing deviations from the perfect sphere, represented by *coefficients* in spherical harmonic series expansions. There are altogether three classes of harmonic expansion coefficients. *Zonal* coefficients, which assume symmetry about the spin axis; *tesseral* ("cube shaped") coefficients; and *sectorial* coefficients, which refer to orange-section-shaped sectors. The expansion into spherical harmonic series is extensively discussed in Baker and Makemson (1967).

Making the basic assumption of rotational symmetry and using satellite data, King-Hele and Cook (1973) have computed the deviations from spheroidal for the geoid. The resulting pear shape is shown in Figure 5.7, superimposed on the Earth spheroid.

One hundred tons of dust from outer space settles on the surface of the Earth every day. It is the debris from the 200 million meteors that strike our planet daily. Most meteors are about the size of a grain of sand, approach the Earth at about 30 km/s, and burn up in the atmosphere. Their contribution to the Earth's mass is so small, that the Earth's mass can generally be considered constant.

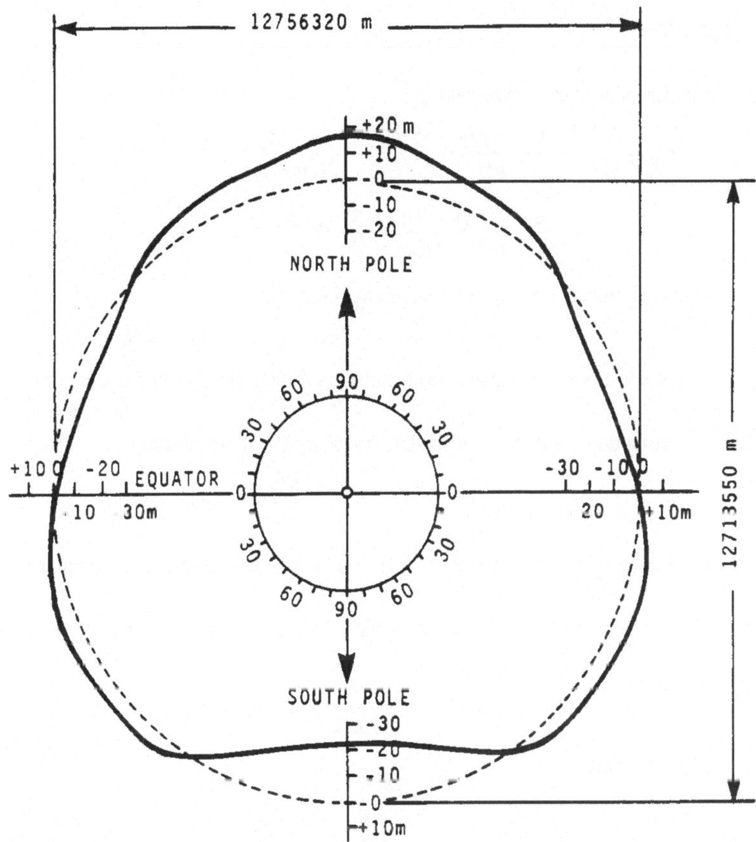

FIGURE 5.7. The geoid's massive shape, as determined from satellite behaviour (according to King-Hele and Cook, 1973).

Suggested Reading

1. Baker, R.; M.W. Makemson. *An Introduction to Astrodynamics*, Academic Press, New York, 1967, 439 pp.

2. Bucerius, H.; M. Schneider. *Himmelsmechanik II*, B.I. Mannheim, 114/144a, 1967, 262 pp.

3. Garland, G.D. *Introduction to Geophysics*, W.B. Saunders Co., 2nd edition, 1979, 494 pp.

4. King-Hele, D.G.; G.E. Cook. *Refining the Earth's Pear Shape*, Nature, 246, 1973, pp. 86-88.

5. Kozai, Y. *Revised Values for Coefficients of Zonal Harmonics in the Geopotential*, Dynamics of Satellites, 1969, (Editor: B. Morando), Springer-Verlag, Berlin, 1970, pp. 104-118.

6. Stacey, F.D. *Physics of the Earth*, John Wiley and Sons, 2nd edition, 1977, 414 pp.

Problems

5.1   The Legendre function of degree n

$$P_n(x) = \frac{1 \cdot 3 \, \cdots \, (2n-1)}{n!}\left\{ x^n - \frac{n(n-1)}{2(2n-1)}x^{n-2} \right.$$

$$\left. + \frac{n(n-1)(n-2)(n-3)}{2 \cdot 4(2n-1)(2n-3)}x^{n-4} - \cdots + \cdots \right\}$$

is the solution of the Legendre differential equation

$$(1-x^2)y'' - 2xy' + n(n+1)y = 0$$

where n is a positive integer. (a) Determine $P_5(x)$. (b) Use Rodrigues' formula to find $P_5(x)$.

5.2   The attraction constant g^* of the Earth spheroid is approximately

$$g^* = \frac{\mu}{r^2}\left\{ 1 - \frac{3}{2}\frac{R_e^2}{r^2}J_2(3\sin^2\delta - 1) \right\}$$

Determine g^* (for the non-rotating (!) Earth) at the equator and at the poles.

5.3   The coefficient of the second harmonic of the zonal function for the gravitational potential of the Earth is defined as

$$J_2 = -\frac{1}{MR_e^2}\int_M r'^2 P_2(\sin\delta')dM$$

Show (a) that

$$J_2 = \frac{C-A}{MR_e^2}$$

and (b) compute the difference between the inertia moments C and A, when

$$J_2 = 0.001\,082\,628 \qquad M = 5.976 \cdot 10^9\ \text{Eg} \qquad R_e = 6\,378.16\,\text{km}$$

(1 Eg = 1 Exagram = 10^{18} g)

5.4   Determine (a) the potential around a master mass consisting of a very thin flat plate of rectangular shape. Also (b) find the attraction force exerted on a satellite of $m = 1$ kg, located at a distance $r = 200$ m, with $\alpha = 30°$ and $\gamma = 60°$, if $M = 6 \cdot 10^{24}$ kg, $a = 100$ m, $b = 50$ m, $c = 1$ m. (*Hint:* Use $\mathbf{F} = -\text{grad}U$ with r as the first coordinate and α as the second coordinate and observe that $\gamma = 90° - \alpha$ for the present problem.) Assume $a^3 \ll r^3$ throughout.

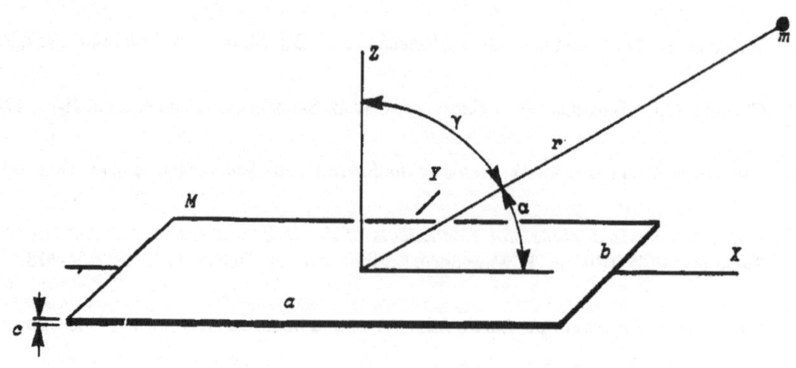

5.5 Given is a master mass M, which has the shape of a torus of radius R with an equally distributed mass. Determine the potential of the master for a small point satellite m located at a distance r. It is $M >> m$ and $R^3 << r^3$. Sketch the equipotential lines in a plane normal to the toroidal plane.

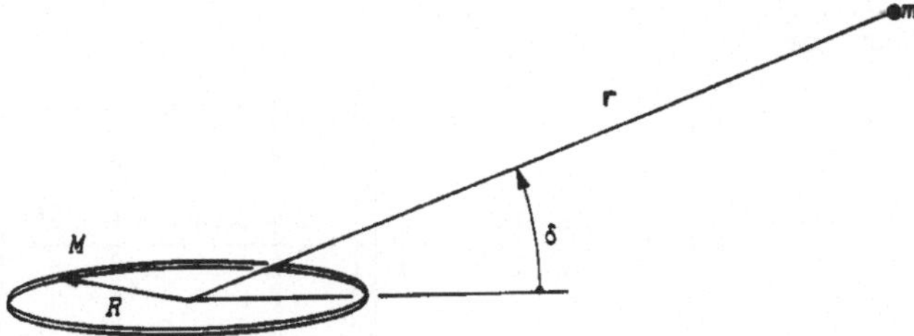

5.6 Given is the master mass M, which has the shape of a thin disk of thickness h and radius R and a uniform mass distribution. (a) Determine the potential surrounding the master mass for a small point satellite of mass m located at a distance r and a declination δ. It is $m << M, h << R, R^3 << r^3$. (b) Sketch the equipotential lines for $r = 2R$ and $r = 3R$ in a plane perpendicular to the disk.

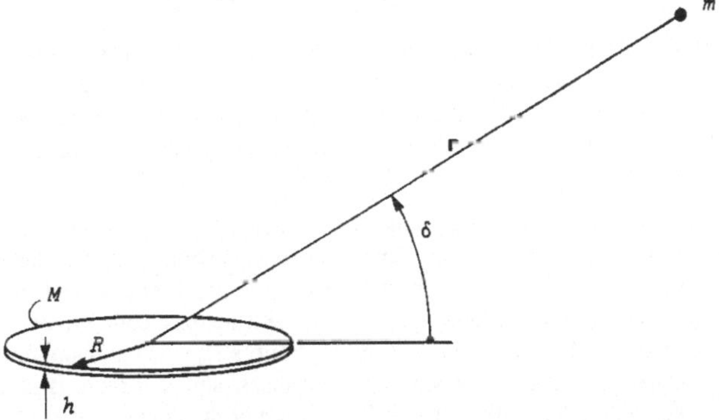

5.7 Given is a symmetrical master M and a point satellite m. The master consists of a solid circular cylinder. Its inner portion, of radius a and height h, has a uniform mass density of 2ρ. Its outer portion, between radii a and b, has a uniform mass density of ρ. The total mass of the master amounts to M. Determine the potential at the location r of the point satellite, taking into consideration terms with, up to and including, the second power of r, for the special case that $a = R/2, b = R$, and $h = R$.

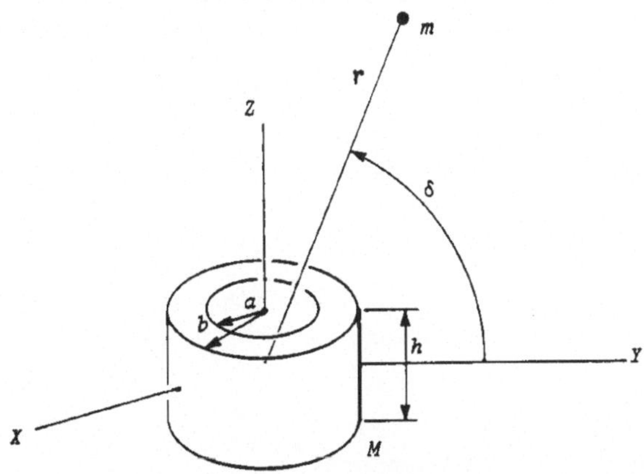

5.8   Given is the potential

$$U = -\frac{\mu m}{\rho_C}\left\{1 + \frac{A + B + C - 3I_C}{2m\rho_C^2}\right\}$$

Show that

$$-\mathbf{grad}\,U = -\frac{\mu m}{\rho_C^2}\left[\mathbf{e}_x\,\mathbf{e}_y\,\mathbf{e}_z\right]\begin{bmatrix}\left\{1 + \frac{3}{2}\frac{3A + B + C - 5I_C}{m\rho_C^2}\right\}\cos\alpha \\[2ex] \left\{1 + \frac{3}{2}\frac{3A + B + C - 5I_C}{m\rho_C^2}\right\}\cos\beta \\[2ex] \left\{1 + \frac{3}{2}\frac{3A + B + C - 5I_C}{m\rho_C^2}\right\}\cos\gamma\end{bmatrix}$$

(*Hint*: Let $\rho_C = \sqrt{x^2 + y^2 + z^2}$ and $\cos\alpha = x/\rho_C$ etc.)

5.9   The Earth's mass is

$$m = 5.976 \cdot 10^{24}\ \text{kg}$$

If 100 tons of dust from meteors settles on the Earth daily, (a) by what percentage will the Earth's mass increase in one year? (b) How many years will it take for the Earth's mass to double? (c) By how much does the Earth's moment of inertia about the North-South axis ($C = 80.416 \cdot 10^{30}$ km^{2}kg) change in a year, if the dust settles uniformly on the surface?

5.10   Due to friction losses brought about by the tidal activity of its oceans, the Earth's angular momentum is continually reduced and transferred to the Moon, such that the combined angular momentum remains constant. Show that if the radius of the Moon's orbit increases by some 4 cm each year, the Earth's day becomes longer by some 2 ms in a century. Make simplifying assumptions, such as that the Moon's orbit is within the Earth's equatorial plane, that the Moon's orbit is circular (384 400 km), that the Moon is a point mass, and that there is no perturbation from the Sun. Also assume that the Earth's axis is the location of the common mass centre. Show that the kinetic energy of the Earth-Moon system declines.

5.11   Show that

$$U = -\frac{\mu m}{r}$$

satisfies the Laplacian differential equation for potential functions

$$\frac{\partial^2 U}{\partial x^2} + \frac{\partial^2 U}{\partial y^2} + \frac{\partial^2 U}{\partial z^2} = 0$$

6

Variation of Geometric Orbital Elements

The term *variation of elements* is not only applied to indicate that the elements of an orbit are affected by small additional forces applied to a satellite (in addition to the Kepler force) and consequently vary, but also indicates a *method* to determine these changes, which was introduced by Euler[*] in 1748.

The method is based on the assumption, that the changes (variations) in the orbital elements are so small, that the elements themselves required to calculate the changes can be considered constant in the calculation.

It is important to realize, that the small variations which we are about to discuss may have an insignificant effect on a satellite orbit. This is indeed so. However, one must not forget that lives of artificial satellites may be of the order of hundreds of years, or much, much longer. During this kind of time, "small" variations may assume impressive magnitudes.

It is also necessary to select a suitable interval, for which the change in a given element is to be calculated. This interval ought to be small enough to insure that the variation is indeed "small". For typical cases, it so happens, that the Kepler period τ turns out to be suitable and, of course, very convenient.

There are several *numerical* methods, in addition to the method of the variation of elements, which are used in practice to determine orbit perturbation, of which the Encke[**] method and the Cowell[***] method are the most important.

The *Encke method* originated in 1857 and is based on a stepwise integration of the difference between the sum of all accelerations and the primary Kepler acceleration. A Kepler orbit serves as reference orbit for each integration step; for the new step a new, corrected Kepler orbit is chosen, with a new position and a new velocity. At the beginning of each step, position and velocity on the reference Kepler orbit and on the perturbed orbit to be computed, coincide. The Encke method is, for instance, well suited for the computation of transfer orbits between Earth and Moon. It is more complex, but faster, than the Cowell method.

The *Cowell method* came into being in 1910. It is a simple method and well suited for orbit computations of many kinds, provided a large digital computer is available. The process involves a step-wise numerical integration of the equations of motion, which include all perturbation accelerations. For large perturbing accelerations, small integration steps must be chosen. Resulting computational inaccuracies restrict the method somewhat.

The perturbation force $\mathbf{f}$ is caused by the perturbation portion (5.39) of the gravitational potential of the master mass. The equations developed in the present chapter are, however, also suitable for the determination of orbital element changes due to other small forces, e.g. magnetic forces, provided they are small in comparison with the Kepler force.

Thus the velocity is seen to change continuously as a result of the perturbation force, in addition to the changes caused by the Kepler force $\mathbf{K}$. For each new velocity, there is a new Kepler orbit with new orbital elements.

Each velocity change, e.g. $\delta v_r = f_r \, \delta t/m$, causes a change of position, e.g. $\delta r = f_r (\delta t)^2/2m$, which is negligibly small compared to the velocity change, since $(\delta t)^2 << \delta t$. In other words, the velocity changes *alone* are the mechanism for the variation of the orbital elements.

In order to determine the rates of change for the five geometric orbital elements, a, ε, Ω, ω, and i, we investigate the influence of a velocity change $\delta v = \mathbf{f} \, \delta t$ (superimposed upon the velocity changes due to the Kepler force) upon the orbital elements, as the satellite is at a position $\mathbf{r} = $ constant. Mathematically speaking, it means, that we consider all quantities constant, except the orbital element whose change we want to calculate.

---

[*] Leonard Euler (1707-1783), Swiss mathematician.

[**] Johann Encke (1791-1865), German astronomer.

[***] Philip Cowell (1870-1949), English astonomer.

6.1   Perturbation Force and Velocity Change

We shall refer to a force **f** which acts on a satellite *in addition* to the Kepler force

$$\mathbf{K} = -\frac{\mu m}{r^2}\, \mathbf{e}_r \tag{6.1}$$

as a *perturbation force*. The perturbation force shall be very much smaller than the Kepler force, i.e.

$$|\mathbf{f}| \ << \ |\mathbf{K}| \tag{6.2}$$

We shall use an orbital coordinate system. The perturbation force thus will have components (Figure 5.6) in radial, transverse, and normal directions, such that

$$\mathbf{f} = \begin{bmatrix} \mathbf{e}_r \, \mathbf{e}_\theta \, \mathbf{e}_z \end{bmatrix} \begin{bmatrix} f_r \\ f_\theta \\ f_z \end{bmatrix} \tag{6.3}$$

The velocity of a satellite on its Kepler orbit is

$$\mathbf{v} = \begin{bmatrix} \mathbf{e}_r \, \mathbf{e}_\theta \, \mathbf{e}_z \end{bmatrix} \begin{bmatrix} v_r \\ v_\theta \\ 0 \end{bmatrix} \tag{6.4}$$

If a perturbation force is acting for a short time interval δt the associated *change* of the velocity will be

$$\delta \mathbf{v} = \begin{bmatrix} \mathbf{e}_r \, \mathbf{e}_\theta \, \mathbf{e}_z \end{bmatrix} \begin{bmatrix} \delta v_r \\ \delta v_\theta \\ \delta v_z \end{bmatrix} = \begin{bmatrix} \mathbf{e}_r \, \mathbf{e}_\theta \, \mathbf{e}_z \end{bmatrix} \begin{bmatrix} \dfrac{f_r}{m}\,\delta t \\[2mm] \dfrac{f_\theta}{m}\,\delta t \\[2mm] \dfrac{f_z}{m}\,\delta t \end{bmatrix} \tag{6.5}$$

6.2   Rate of Change of Magnitude of Semi-Major Axis

A point satellite on an elliptical orbit has a velocity, whose magnitude is governed by the speed equation (2.13)

$$v = \sqrt{\mu\left[\frac{2}{r} - \frac{1}{a}\right]} \tag{6.6}$$

The speed equation can be rearranged such that it provides an expression for the semi-major axis and allows for the fact that the velocity v has a radial and a transverse component

$$a = \frac{\mu r}{2\mu - rv_r^2 - rv_\theta^2} \tag{6.7}$$

We now find the variation δa by forming a total differential

$$\delta a = \frac{\partial a}{\partial r}\,\delta r + \frac{\partial a}{\partial \mu}\,\delta \mu + \frac{\partial a}{\partial v_r}\,\delta v_r + \frac{\partial a}{\partial v_\theta}\,\delta v_\theta \tag{6.8}$$

and make the following assumptions:

1.   The satellite is located (during the *short* time interval δt) at r = constant.
2.   The satellite is located in the field of *one* master, i.e. μ = constant.

3. Any deviations from a pure Kepler case are due to a perturbation force (6.3) which acts on the satellite at position **r** in addition to the Kepler force.

Observing that $\delta r = 0$ because of assumption 1, and that $\delta\mu = 0$ because of assumption 2, there remains

$$\delta a = \frac{\partial a}{\partial v_r}\, \delta v_r + \frac{\partial a}{\partial v_\theta}\, \delta v_\theta \tag{6.9}$$

At the position $r = $ constant the satellite switches instantaneously from a Kepler orbit of semi-major axis a to an osculating Kepler orbit of semi-major axis $a + \delta a$.

We now form the partial derivatives $\partial a/\partial v_r$ and $\partial a/\partial v_\theta$ from equation (6.7) and obtain for equation (6.9)

$$\delta a = \frac{2a^2 v_r}{\mu}\, \delta v_r + \frac{2a^2 v_\theta}{\mu}\, \delta v_\theta \tag{6.10}$$

In order to obtain the variation in a as function of the true anomaly we express v_r and v_θ as functions of θ. This can be done with the aid of equations (2.6) and (2.7), giving

$$v_r = \dot{r} = \sqrt{\frac{\mu}{p}}\, \varepsilon \sin\theta \tag{6.11}$$

and

$$v_\theta = r\dot\theta = \sqrt{\frac{\mu}{p}} = \sqrt{\frac{\mu}{p}}\, (1 + \varepsilon \cos\theta) \tag{6.12}$$

Collecting terms, observing that $\delta v_r = f_r \delta t / m$ and $\delta v_\theta = f_\theta \delta t / m$ from equation (6.5), and introducing da/dt instead of $\delta a/\delta t$, the rate of change of the semi-major axis is obtained as

$$\frac{da}{dt} = \sqrt{\frac{p}{\mu}}\, \frac{2p}{(1-\varepsilon^2)^2} \left[\varepsilon \sin\theta\, \frac{f_r}{m} + (1 + \varepsilon \cos\theta)\, \frac{f_\theta}{m} \right] \tag{6.13}$$

Equation (6.13) is listed in Table 6.1.

If one now wishes to introduce the true anomaly θ instead of the time t, then

$$\frac{da}{d\theta} = \frac{da}{dt}\, \frac{dt}{d\theta} = \frac{1}{\dot\theta}\, \frac{da}{dt} \tag{6.14}$$

and since according to the constancy of the angular momentum as e.g. according to equation (2.2), which is considered to be essentially valid (because the perturbation force is very small),

$$\dot\theta = \sqrt{\frac{\mu}{p^3}}\, (1 + \varepsilon \cos\theta)^2 \tag{6.15}$$

we obtain eventually

$$\frac{da}{d\theta} = \frac{p^2}{\mu}\, \frac{2p}{(1-\varepsilon^2)^2} \left[\frac{f_r}{m} + \frac{1}{1 + \varepsilon \cos\theta}\, \frac{f_\theta}{m} \right] \tag{6.16}$$

Equation (6.16) is shown again in Table 6.2.

6.3 Rate of Change of Eccentricity

Before studying the eccentricity, let us first have a look at the semi-parameter p. Constancy of angular momentum requires that

$$r^2 \dot\theta = h = \sqrt{\mu p} \tag{6.17}$$

With $v_\theta = r\dot\theta$ and solved for p,

$$p = \frac{r^2 v_\theta^2}{\mu} \tag{6.18}$$

The total differential is

$$dp = \frac{\partial p}{\partial r}\, dr + \frac{\partial p}{\partial \mu}\, d\mu + \frac{\partial p}{\partial v_\theta}\, dv_\theta \tag{6.19}$$

Using again the conditions $dr = 0$ and $d\mu = 0$, one can conclude that the semi-parameter will change as a result of dv_θ only.

$$dp = \frac{\partial p}{\partial v_\theta}\, dv_\theta \tag{6.20}$$

Partial differentiation of equation (6.18), and use of $dv_\theta = (f_\theta/m)dt$ and $v_\theta = \sqrt{\mu p / r}$, gives us the time rate of change of the semi-parameter caused by the perturbation force,

$$\dot{p} = \frac{dp}{dt} = 2r \sqrt{\frac{p}{\mu}}\, \frac{f_\theta}{m} \tag{6.21}$$

We are now in a position to determine the rate of change of the eccentricity ε. Semi-major axis, semi-parameter, and eccentricity are related (Table 2.1) by

$$p = a(1 - \varepsilon^2)$$

Differentiation with respect to time gives

$$\dot{p} = \dot{a}(1 - \varepsilon^2) - 2a\varepsilon\dot{\varepsilon} \tag{6.22}$$

With $\dot{p}$ from equation (6.21) and $\dot{a}$ from (6.14), and by solving for $\dot{\varepsilon}$, the time rate of change of the eccentricity becomes

$$\dot{\varepsilon} = \frac{d\varepsilon}{dt} = \frac{1}{m} \sqrt{\frac{p}{\mu}} \left[\sin\theta\, f_r + \frac{1}{\varepsilon}\left[\frac{p}{r} - \frac{r}{a}\right] f_\theta \right] \tag{6.23}$$

or, using the true anomaly exclusively,

$$\dot{\varepsilon} = \frac{d\varepsilon}{dt} = \sqrt{\frac{p}{\mu}} \left[\sin\theta\, \frac{f_r}{m} + \frac{\varepsilon + 2\cos\theta + \varepsilon\cos^2\theta}{1 + \varepsilon\cos\theta}\, \frac{f_\theta}{m} \right] \tag{6.24}$$

Equation (6.24) is also listed in Table 6.1.

The rate of change of the eccentricity can also be expressed by using the true anomaly as independent variable, with the help of equations (6.15) and (6.16).

The result is

$$\frac{d\varepsilon}{d\theta} = \frac{p^2}{\mu} \left[\frac{\sin\theta}{(1 + \varepsilon\cos\theta)^2}\, \frac{f_r}{m} + \frac{\varepsilon + 2\cos\theta + \varepsilon\cos^2\theta}{(1 + \varepsilon\cos\theta)^3}\, \frac{f_\theta}{m} \right] \tag{6.25}$$

Equation (6.25) is also listed in Table 6.2.

6.4 Rate of Change of Right Ascension of Ascending Node

An inspection of Figure 6.1 will indicate that a velocity change dv_z of the satellite m perpendicular to the orbital plane is associated with a tilt $d\kappa$ of the orbital plane about the satellite's position vector $\mathbf{r}$ as tilt axis. From Figure 6.1 one concludes that

$$d\kappa = \frac{dv_z}{r\dot{\theta}} \tag{6.26}$$

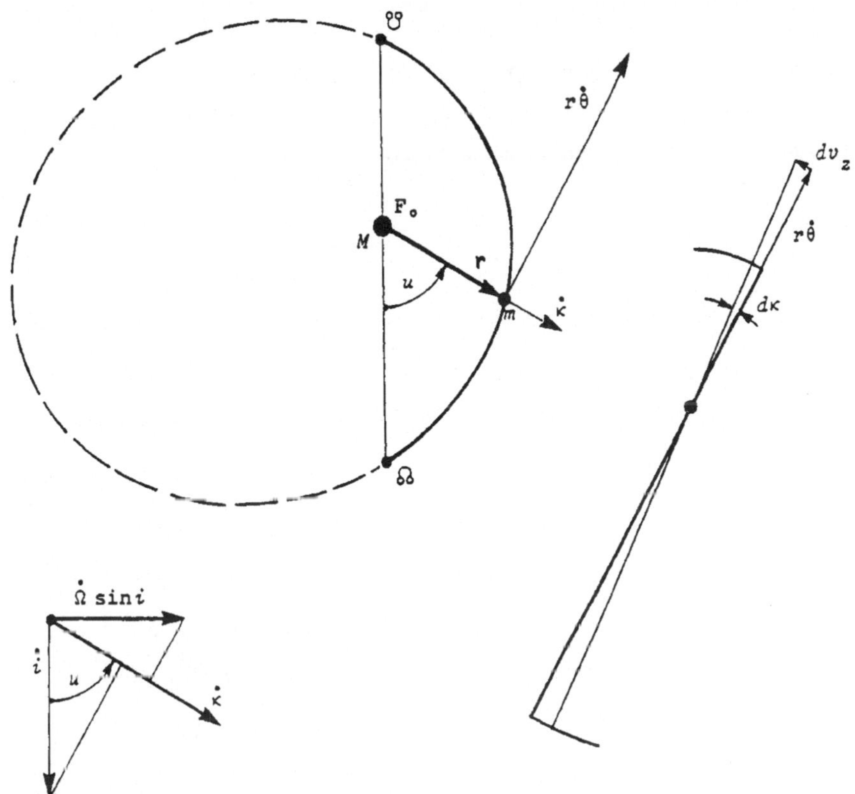

FIGURE 6.1. Precession $\dot{\Omega}$ and inclination change $\dot{i}$ both contribute to $\dot{\kappa}$.

and since $dv_z = f_z\,dt/m$, one obtains for the angular velocity of the orbital plane about the radius vector

$$\dot{\kappa} = \frac{1}{r\dot{\theta}}\frac{f_z}{m} \tag{6.27}$$

Both, the rate $\dot{\Omega}$ of change of the right ascension of the ascending node, and the rate $\dot{i}$ of change of the orbital plane inclination angle contribute to the tilt rate $\dot{\kappa}$ (Figure 6.2)

$$\dot{\kappa} = \dot{i}\cos u + \dot{\Omega}\sin i \sin u \tag{6.28}$$

The component perpendicular to $\dot{\kappa}$ must, of course, vanish (if dv_z is the only velocity change),

$$\dot{i}\sin u - \dot{\Omega}\sin i \cos u = 0 \tag{6.29}$$

Equations (6.27), (6.28), and (6.29), rearranged, and considering that $u = \theta + \omega$, give for the rate of change of the orientation of the line of nodes

$$\dot{\Omega} = \frac{1}{r\dot{\theta}}\frac{\sin(\theta+\omega)}{\sin i}\frac{f_z}{m} = \sqrt{\frac{p}{\mu}}\frac{1}{\sin i}\frac{\sin(\theta+\omega)}{1+\varepsilon\cos\theta}\frac{f_z}{m} \tag{6.30}$$

Using equations (6.15) and (6.16) again, we obtain an angular rate of change of the orientation of the line of nodes

$$\frac{d\Omega}{d\theta} = \frac{p^2}{\mu} \frac{1}{\sin i} \frac{\sin(\theta + \omega)}{(1 + \varepsilon \cos \theta)^3} \frac{f_z}{m} \tag{6.31}$$

Equation (6.30) is listed in Table 6.1 and equation (6.31) in Table 6.2.

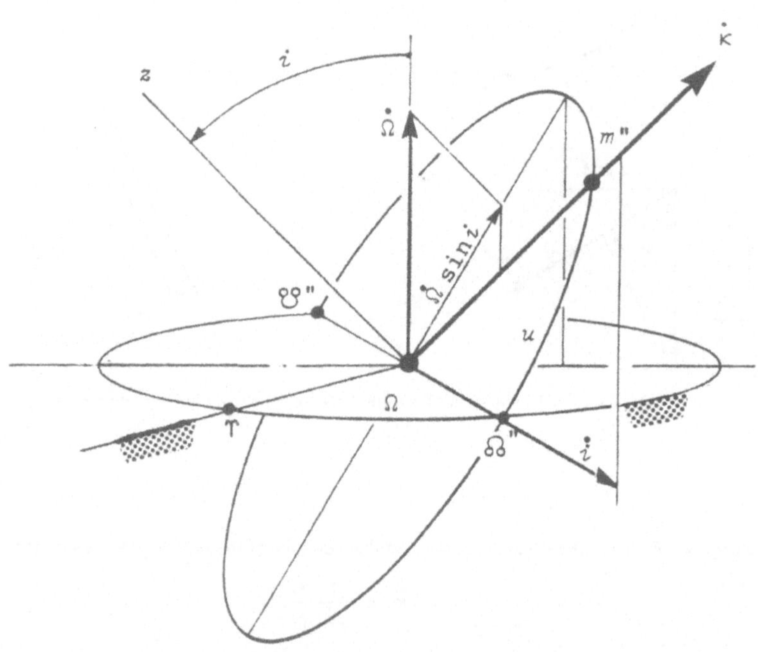

FIGURE 6.2. The angular velocities $\dot{\Omega}$, $\dot{i}$, and $\dot{\kappa}$.

6.5   Rate of Change of Orientation of Line of Apsides

The orientation of the line of apsides is given by the orbital element ω (Figure 2.13). In place of the angle ω we introduce temporarily the angle ω', which gives the absolute orientation of the periapsis with respect to an (imagined) fixed reference system (Figure 6.3). We now pose the question as to any change of the angle ω' as a result of the action of a perturbation force $\mathbf{f}$.

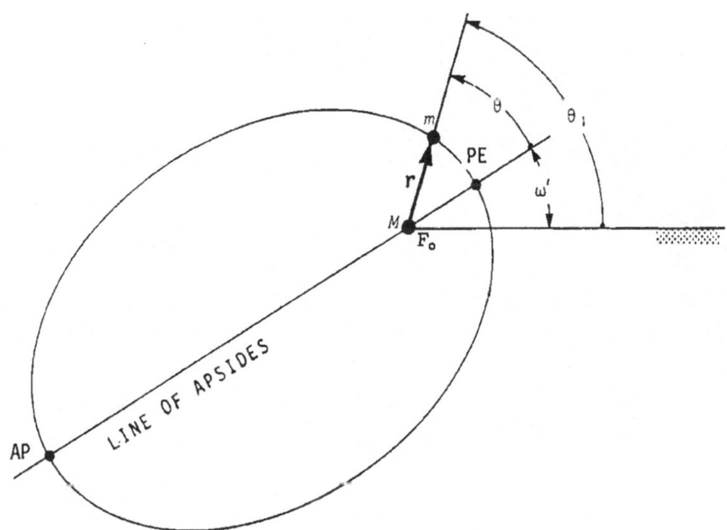

FIGURE 6.3. Absolute orientation of the line of apsides.

We use the equation for a Kepler ellipse in the form (Figure 6.3)

$$r(1 + \varepsilon \cos(\theta_1 - \omega')) = p \tag{6.32}$$

and investigate the change of the orbital elements at $r = $ constant and $\theta_1 = $ constant.

$$r\dot{\varepsilon} \cos(\theta_1 - \omega') - r\varepsilon \sin(\theta_1 - \omega')\dot{\omega}' = \dot{p} \tag{6.33}$$

We know from equation (6.24) that

$$\dot{\varepsilon} = \sqrt{\frac{p}{\mu}} \left[\sin\theta \, \frac{f_r}{m} + \frac{\varepsilon + 2\cos + \varepsilon \cos^2\theta}{1 + \varepsilon \cos\theta} \, \frac{f_\theta}{m} \right] \tag{6.34}$$

and from equation (6.21) we know that

$$\dot{p} = \sqrt{\frac{p}{\mu}} \, \frac{2p}{1 + \varepsilon \cos\theta} \, \frac{f_\theta}{m} \tag{6.35}$$

We combine and obtain

$$\dot{\omega}' = \frac{d\omega'}{dt} = \sqrt{\frac{p}{\mu}} \left[-\frac{\cos\theta}{\varepsilon} \, \frac{f_r}{m} + \frac{(2 + \varepsilon \cos\theta)\sin\theta}{\varepsilon(1 + \varepsilon \cos\theta)} \, \frac{f_\theta}{m} \right] \tag{6.36}$$

The quantity $\dot{\omega}'$ represents an absolute angular velocity, i.e. the velocity at which the line of apsides rotates with respect to an inertially fixed coordinate system. As orbital element we would like to use, however, the angle ω between line of apsides and line of nodes. The line of nodes itself changes at an absolute rate $\dot{\Omega}$, whose component normal to the orbital plane is

$$\dot{\Omega} \cos i \tag{6.37}$$

Consequently (Figure 6.4)

$$\dot{\omega} = \dot{\omega}' - \dot{\Omega} \cos i \tag{6.38}$$

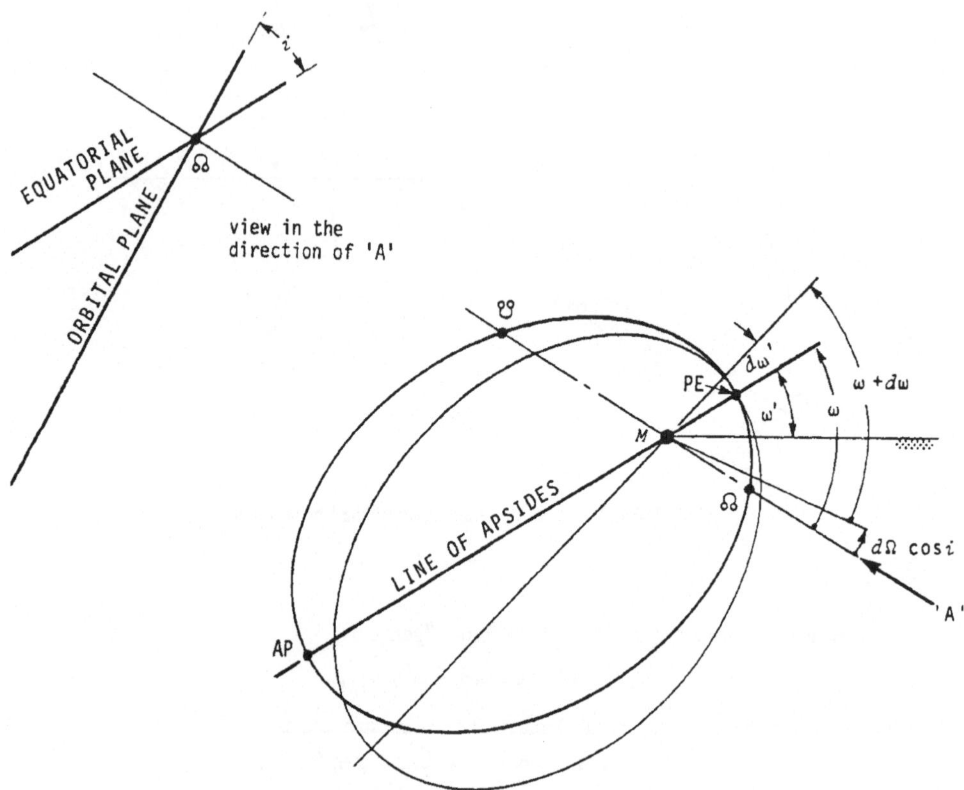

FIGURE 6.4. Absolute and relative rotation of the line of apsides with respect to the line of nodes.

The precession $\dot{\Omega}$ of the orbital plane is known from equation (6.30). Eventually for the time rate of change of the orientation of the line of apsides one arrives at

$$\dot{\omega} = \frac{d\omega}{dt} = \sqrt{\frac{p}{\mu}} \left[-\frac{\cos\theta}{\varepsilon} \frac{f_r}{m} + \frac{(2 + \varepsilon\cos\theta)\sin\theta}{\varepsilon(1 + \varepsilon\cos\theta)} \frac{f_\theta}{m} - \frac{1}{\tan i} \frac{\sin(\theta + \omega)}{1 + \varepsilon\cos\theta} \frac{f_z}{m} \right] \tag{6.39}$$

Multiplication with $1/\dot{\theta}$ gives the rate of change of the orientation of the line of apsides with respect to the true anomaly

$$\frac{d\omega}{d\theta} = \frac{p^2}{\mu} \left[-\frac{\cos\theta}{\varepsilon(1 + \varepsilon\cos\theta)^2} \frac{f_r}{m} + \frac{(2 + \varepsilon\cos\theta)\sin\theta}{\varepsilon(1 + \varepsilon\cos\theta)^3} \frac{f_\theta}{m} - \frac{1}{\tan i} \frac{\sin(\theta + \omega)}{(1 + \varepsilon\cos\theta)^3} \frac{f_z}{m} \right] \tag{6.40}$$

Equations (6.39) and (6.40) are listed in Tables 6.1 and 6.2 respectively.

$$\frac{da}{dt} = \sqrt{\frac{p}{\mu}} \, \frac{2p}{(1-\varepsilon^2)^2} \left[\varepsilon \sin\theta \, \frac{f_r}{m} + (1 + \varepsilon\cos\theta) \, \frac{f_\theta}{m} \right] \qquad (6.13)$$

$$\frac{d\varepsilon}{dt} = \sqrt{\frac{p}{\mu}} \left[\sin\theta \, \frac{f_r}{m} + \frac{\varepsilon + 2\cos\theta + \varepsilon\cos^2\theta}{1 + \varepsilon\cos\theta} \, \frac{f_\theta}{m} \right] \qquad (6.24)$$

$$\frac{d\Omega}{dt} = \sqrt{\frac{p}{\mu}} \, \frac{1}{\sin i} \, \frac{\sin(\theta+\omega)}{1 + \varepsilon\cos\theta} \, \frac{f_z}{m} \qquad (6.30)$$

$$\frac{d\omega}{dt} = \sqrt{\frac{p}{\mu}} \left[-\frac{\cos\theta}{\varepsilon} \, \frac{f_r}{m} + \frac{(2 + \varepsilon\cos\theta)\sin\theta}{\varepsilon(1 + \varepsilon\cos\theta)} \, \frac{f_\theta}{m} \right.$$
$$\left. - \frac{1}{\tan i} \, \frac{\sin(\theta + \omega)}{1 + \varepsilon\cos\theta} \, \frac{f_z}{m} \right] \qquad (6.39)$$

$$\frac{di}{dt} = \sqrt{\frac{p}{\mu}} \, \frac{\cos(\theta + \omega)}{1 + \varepsilon\cos\theta} \, \frac{f_z}{m} \qquad (6.42)$$

TABLE 6.1. Time rates of change of orbital elements due to a small perturbation force.

6.6 Rate of Change of Inclination

The time rate of change of the inclination i of the orbital plane is derived from equations (6.29) and (6.30), and is

$$\dot{i} = \frac{di}{dt} = \frac{1}{r\dot{\theta}} \cos(\theta + \omega) \, \frac{f_z}{m} \qquad (6.41)$$

or

$$\dot{i} = \frac{di}{dt} = \sqrt{\frac{p}{\mu}} \, \frac{r}{p} \cos(\theta + \omega) \, \frac{f_z}{m} = \sqrt{\frac{p}{\mu}} \, \frac{\cos(\theta + \omega)}{1 + \varepsilon\cos\theta} \, \frac{f_z}{m} \qquad (6.42)$$

The derivative with respect to the true anomaly is obtained in the same way as in the previous cases

$$\frac{di}{d\theta} = \frac{p^2}{\mu} \, \frac{\cos(\theta + \omega)}{(1 + \varepsilon\cos\theta)^3} \, \frac{f_z}{m} \qquad (6.43)$$

Equations (6.42) and (6.43) are listed again in Tables 6.1 and 6.2 respectively.

$$\frac{da}{d\theta} = \frac{p^2}{\mu} \frac{2p}{(1-\varepsilon^2)^2} \left[\frac{\varepsilon \sin \theta}{(1+\varepsilon \cos \theta)^2} \frac{f_r}{m} + \frac{1}{1+\varepsilon \cos \theta} \frac{f_\theta}{m} \right] \qquad (6.16)$$

$$\frac{d\varepsilon}{d\theta} = \frac{p^2}{\mu} \left[\frac{\sin \theta}{(1+\varepsilon \cos \theta)^2} \frac{f_r}{m} + \frac{\varepsilon + 2\cos \theta + \varepsilon \cos^2 \theta}{(1+\varepsilon \cos \theta)^3} \frac{f_\theta}{m} \right] \qquad (6.25)$$

$$\frac{d\Omega}{d\theta} = \frac{p^2}{\mu} \frac{1}{\sin i} \frac{\sin(\theta+\omega)}{(1+\varepsilon \cos \theta)^3} \frac{f_z}{m} \qquad (6.31)$$

$$\frac{d\omega}{d\theta} = \frac{p^2}{\mu} \left[-\frac{\cos\theta}{\varepsilon(1+\varepsilon \cos \theta)^2} \frac{f_r}{m} + \frac{(2+\varepsilon \cos \theta)\sin \theta}{\varepsilon(1+\varepsilon \cos \theta)^3} \frac{f_\theta}{m} \right.$$
$$\left. - \frac{1}{\tan i} \frac{\sin(\theta + \omega)}{(1+\varepsilon \cos \theta)^3} \frac{f_z}{m} \right] \qquad (6.40)$$

$$\frac{di}{d\theta} = \frac{p^2}{\mu} \frac{\cos (\theta + \omega)}{(1+\varepsilon \cos \theta)^3} \frac{f_z}{m} \qquad (6.43)$$

TABLE 6.2. Derivatives of orbital elements with respect to true anomaly.

Problems

6.1   Given is the time rate of change of the orbital element a,

$$\frac{da}{dt} = \sqrt{\frac{p}{\mu}} \frac{2a}{1-\varepsilon^2} \left\{ \varepsilon \sin \theta \frac{f_r}{m} + (1+\varepsilon \cos \theta) \frac{f_\theta}{m} \right\}$$

Show by means of a transformation of coordinates that

$$\frac{da}{dt} = \frac{2a^2 v}{\mu} \frac{f_t}{m}$$

where

$$\mathbf{f} = \left[\mathbf{e}_r \, \mathbf{e}_\theta \, \mathbf{e}_z \right] \begin{bmatrix} f_r \\ f_\theta \\ f_z \end{bmatrix} = \left[\mathbf{e}_n \, \mathbf{e}_t \, \mathbf{e}_z \right] \begin{bmatrix} f_n \\ f_t \\ f_z \end{bmatrix}$$

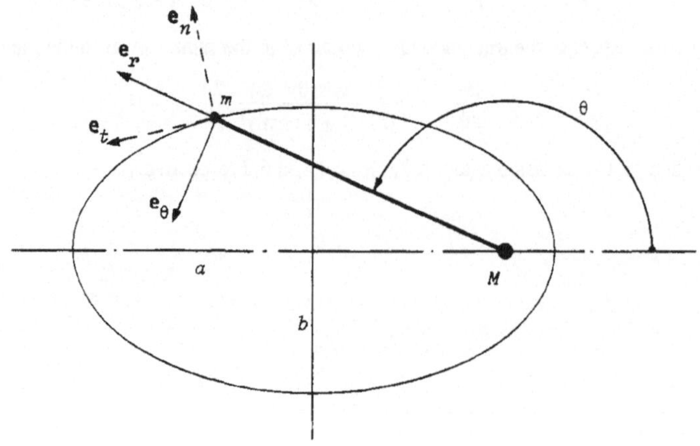

6.2   A perturbation force

$$\mathbf{f} = \begin{bmatrix} \mathbf{e}_r & \mathbf{e}_\theta & \mathbf{e}_z \end{bmatrix} \begin{bmatrix} 0 \\ 0 \\ 2\cos\theta \end{bmatrix} \text{ N}$$

is acting on a satellite ($m = 500$ kg) in a LEO (low earth orbit) of $r = 6570$ km, $\varepsilon = 0$, $i = 30°$, for the duration of one orbit. What is the new inclination angle i?

6.3   A constant perturbation force

$$\mathbf{f} = \begin{bmatrix} \mathbf{e}_r & \mathbf{e}_\theta & \mathbf{e}_z \end{bmatrix} \begin{bmatrix} 0 \\ 0.3 \\ 0 \end{bmatrix} \text{ N}$$

is acting on an Earth satellite of 400 kg mass, on a circular orbit of $p = 6600$ km, $\varepsilon = 0$, $i = 10°$. Calculate by how much each geometrical orbital element changes, per orbit.

6.4   While passing through the perigee ($\theta = 0$) a very short impulse

$$\mathbf{f}\Delta t = \begin{bmatrix} \mathbf{e}_r & \mathbf{e}_\theta & \mathbf{e}_z \end{bmatrix} \begin{bmatrix} 0 \\ 0 \\ 80 \end{bmatrix} \text{ kN s}$$

is applied to a satellite of 300 kg mass. The original orbit is characterized by $p = 9500$ km, $\varepsilon = 0.4$, $\omega = 45°$ and $i = 30°$. By how much does each of the five geometrical orbital elements change as a result of the impulse?

6.5   While passing through the apogee ($\theta = 90°$) a very short impulse

$$\mathbf{f}\Delta t = \begin{bmatrix} \mathbf{e}_r & \mathbf{e}_\theta & \mathbf{e}_z \end{bmatrix} \begin{bmatrix} 0 \\ 0 \\ 80 \end{bmatrix} \text{ kN s}$$

is applied to a satellite of 300 kg mass. The original orbit is characterized by $p = 9500$ km, $\varepsilon = 0.4$, $\omega = 45°$, and $i = 30°$. By how much does each of the five geometrical orbit elements change as a result of the impulse?

6.6   While passing through the perigee ($\theta = 0°$) a perturbation force

$$\mathbf{f} = \begin{bmatrix} \mathbf{e}_r & \mathbf{e}_\theta & \mathbf{e}_z \end{bmatrix} \begin{bmatrix} 0 \\ 0 \\ 5 \end{bmatrix} \text{ kN}$$

is acting while the satellite ($m = 200$ kg) traverses an arc of 0.5° of true anomaly ($\Delta\theta = 0.5°$). By how much are the five geometric orbit elements affected, if $p = 9500$ km, $\varepsilon = 0.4$, $\omega = 45°$, and $i = 30°$?

6.7   While passing the point $\theta = 120°$ on its orbit ($p = 9500$ km, $\varepsilon = 0.4$, $\omega = 45°$, $i = 30°$), a satellite's ($m = 250$ kg) thruster system accidentally comes on for a very short time, imparting an impulse of

$$\mathbf{f}\Delta t = \begin{bmatrix} 12 \\ 4 \\ 3 \end{bmatrix} \text{ kN s}$$

to the satellite. By how much are its five geometrical orbit elements affected?

6.8   Given is the Kepler ellipse of an Earth satellite. A small constant transverse perturbation force f_θ is acting on the satellite. (a) Determine the change Δa of the semi-major axis for one orbit; (b) Compute Δa for a satellite of 500 kg mass, if $a = 13\,000$ km, $\varepsilon = 0.5$, $f_\theta = 0.2$ N.

6.9   Given is the time rate of change of the semi-major axis of a Kepler orbit, when the satellite is subjected to a small tangential perturbation force f_t,

$$\frac{da}{dt} = \frac{2a^2 v}{\mu} \frac{f_t}{m}$$

(a) Derive an expression for $da/d\theta$ as function of θ (b) For a constant tangential force f_t and an (originally) circular orbit, how much is the increase Δa for one orbit? Sketch a 5 orbit path of the satellite, and (c) calculate Δa for $\varepsilon = 0$, $a = 10\,000$ km, $f_t = 0.4$ N, and $m = 500$ kg.

6.10 At the moment when a satellite ($m = 150$ kg) of the Earth passes the true anomaly $\theta = 90°$ on its orbit ($r_{PE} = 6600$ km, $\varepsilon = 0.2$, $i = 15°$, $\omega = 30°$), it is given a small impulse

$$\mathbf{f}\Delta t = \begin{bmatrix} \mathbf{e}_r \, \mathbf{e}_\theta \, \mathbf{e}_z \end{bmatrix} \begin{bmatrix} 0.2 \\ 0.1 \\ 0.3 \end{bmatrix} \text{ kN s}$$

of very short duration. What influence has this impulse on each of the five geometric orbital elements?

6.11 A spacecraft ($m = 1000$ kg) orbits the Earth on an elliptic orbit, with $\varepsilon = 0.5$ and $r_{PE} = 6600$ km. At the moment of periapsis passage, a carefully controlled thruster system is turned on and provides a force, over and above the Kepler force, of

$$\mathbf{f} = \begin{bmatrix} \mathbf{e}_r \, \mathbf{e}_\theta \, \mathbf{e}_z \end{bmatrix} \begin{bmatrix} 0 \\ 15(1 + 0.5 \cos \theta)^3 \\ 0 \end{bmatrix} \text{ mN}$$

The thruster system is turned off again after the completion of one orbit. Assume that mass losses are negligible. (a) Sketch the orbit and show $\mathbf{f}$ for $\theta = 0°$, $90°$, $180°$, $270°$. (b) What is the new orbit eccentricity?

7

Secular Variations of the Orbital Elements

If a point satellite is orbiting a master, whose mass distribution is not perfectly spherical, then the elements of the satellite orbit are subject to variations, which depend upon the shape of the master mass. These variations are small when compared to the orbital elements themselves.

Some variations are referred to as periodic, others as secular. Variations which have a positive value at one time, and subsequently a negative value, such that their sum vanishes over one period are called *periodic variations*. Periodic variations of *small* duration are those, whose period is equal to one orbital period, or integral fractions of one orbital period. Periodic variations of *long* duration are those, whose period is equal to one revolution of the line of apsides, or integral fractions thereof.

Variations whose sum does not vanish over one period are called *secular*. Over one orbital period, such secular variations are typically very small. Over longer times, however, they lead to considerable deviations from an ideal Kepler orbit.

Subsequently we shall derive formulas for the secular variations of orbital elements of a point satellite in a gravitational potential

$$U = -\frac{\mu m}{r}\left[1 - \frac{1}{2}J_2\,\frac{R_e^2}{r^2}\,(3\sin^2\delta - 1)\right] \tag{7.1}$$

which is known to approximate conditions for Earth satellites very closely.

7.1   Precession of the Orbital Plane

Because of the influence of the equatorial bulge of the Earth, as represented by equation (7.1), the orbital plane of an Earth satellite is revolving slowing. This effect is called the *precession of the orbital plane*. In Figure 7.1 the (exaggerated) equatorial bulge of the Earth is shown. During its dwell in the northern hemisphere, a satellite is more strongly attracted by bulge II than by bulge III, while during its dwell in the southern hemisphere, it is more strongly attracted by bulge III than by bulge II.

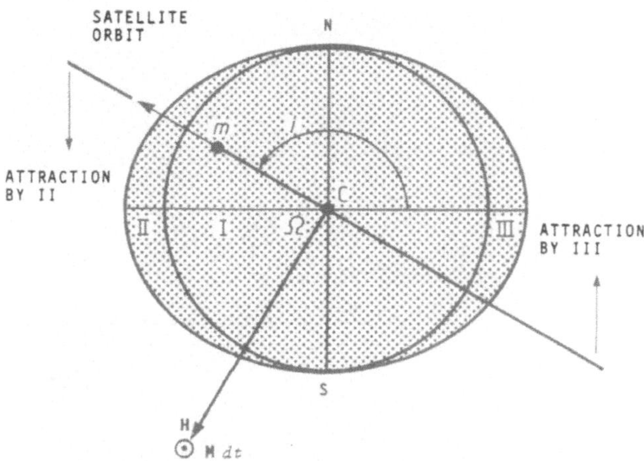

FIGURE 7.1. Earth spheroid and satellite.

As a consequence of the presence of the perturbation force

$$\mathbf{f} \;=\; \begin{bmatrix} \mathbf{e}_r & \mathbf{e}_\theta & \mathbf{e}_z \end{bmatrix} \begin{bmatrix} f_r \\ f_\theta \\ f_z \end{bmatrix} \tag{7.2}$$

which acts on the satellite, the right ascension of Ω of the ascending node (Figure 7.2) varies according to equation (6.31). The component f_z of the perturbation force is given by equation (5.48).

Combining equations (6.31) and (5.48) leads to

$$\frac{d\Omega}{d\theta} \;=\; -\frac{3J_2 R_e^2}{p^2}\,(\sin^2(\theta + \omega))(1 + \varepsilon \cos\theta)\cos i \tag{7.3}$$

By integrating equation (7.3), one obtains a general expression for the orientation Ω of the nodal line as function of the satellite's true anomaly θ. This expression contains periodic as well as secular terms. The periodic terms vanish, if we choose one whole orbital period as integration interval. This we shall do and we thus obtain the secular variation $\Delta\Omega$ of the nodal line for one orbital period.

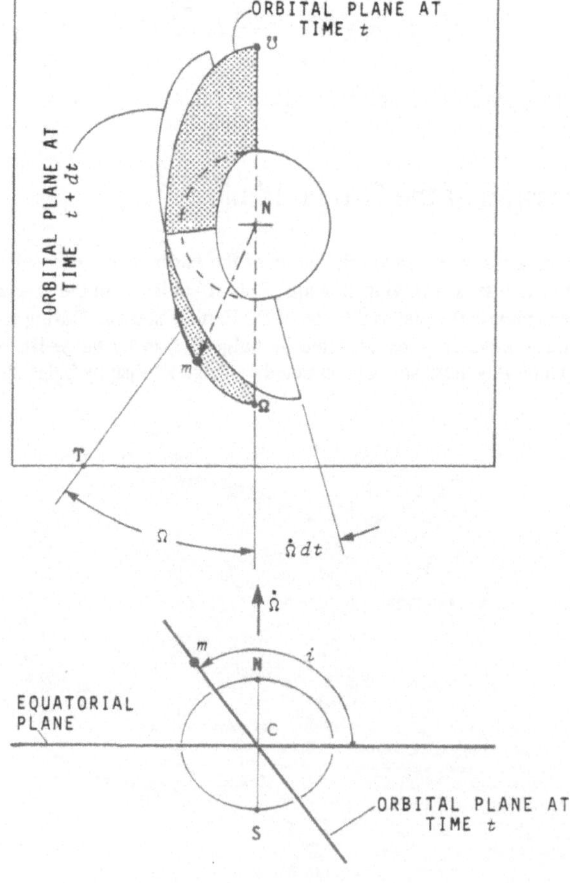

FIGURE 7.2. Precession of the orbital plane.

$$\int_0^{\Delta\Omega} d\Omega \;=\; -\frac{3J_2 R_e^2 \cos i}{p^2} \int_0^{2\pi} (\sin^2(\theta + \omega))(1 + \varepsilon \cos \theta)\, d\theta \tag{7.4}$$

resulting in

$$\Delta\Omega \;=\; -\frac{3\pi J_2 R_e^2}{p^2}\cos i \tag{7.5}$$

The minus sign indicates that the precession of the orbital plane is *retrograde* as long as the inclination of the orbit is less than 90°. For polar orbits ($i = 90°$), there is no precession of the orbital plane. The occurrence of the semi-parameter p in the denominator is an indication that the precession is more pronounced for satellites closer to Earth.

The smaller the angle i, between the satellite's orbital plane and the equatorial plane, the greater is the precession of the orbital plane.

With $J_2 \;=\; 1082.628 \cdot 10^{-6}$, we can write equation (7.5) also as

$$\Delta\Omega \;=\; -0.584\,619 \frac{R_e^2 \cos i}{a^2(1 - \varepsilon^2)^2}\ \text{degree} \tag{7.6}$$

and with the aid of the Kepler period $\tau \;=\; 2\pi\sqrt{a^3/\mu}$ and with $\mu \;=\; 398\,601.19\ \text{km}^3/\text{s}^2$ and $R_e \;=\; 6378.16$ km, the *secular precession of the orbital plane* can be written

$$\bar{\dot\Omega} \;=\; \frac{\Delta\Omega}{\tau} \;=\; -9.964 \left[\frac{R_e}{a}\right]^{7/2} \frac{\cos i}{(1 - \varepsilon^2)^2}\ \frac{\text{degree}}{\text{day}} \tag{7.7}$$

from which it can be seen, that a satellite near the Earth ($a \approx R_e$, $\varepsilon = 0$) on a near equatorial orbit ($i \approx 0°$) has a retrograde precession of its orbital plane of almost $10°$ per day. (For exact equatorial orbits ($i = 0°$) the term "ascending node" is no longer defined.) The quantity $\bar{\dot\Omega}$ represents an *average* angular velocity, in contradistinction to the instantaneous angular velocity $\dot\Omega$ of equation (6.30).

The orbital plane of a satellite of the Earth is *sun-synchronous* if the orbital plane precesses as fast as the apparent precession of the Sun, viz. at

$$\frac{\Delta\Omega}{\tau} \;=\; \frac{360°}{365.2422} \;=\; 0.985\,647\ \frac{\text{degree}}{\text{day}}$$

With $a \approx R_e$, $\varepsilon = 0$, this means an inclination angle i obtainable from

$$\cos i \;=\; -\frac{0.985\,647}{9.964} \;=\; -0.098\,921$$

i.e.

$$i \;=\; 95.677°$$

7.2   Rotation of the Line of Apsides

The line connecting the periapsis and the apoapsis of an elliptic orbit is the *line of apsides*. It passes through the mass centre C of the master mass. The orientation of the line of apsides with respect to the nodal line $\mho$ – C – Ω is given by the argument ω of periapsis, i.e.

$$\omega \;=\; \measuredangle\,(PE - C - \Omega) \tag{7.8}$$

as shown in Figure 7.3. When a perturbation force is acting, the variation of the argument ω is given by equation (6.40). With a perturbation force, whose components are given by equations (5.46), (5.47) and (5.48), and subsequent integration of equation (6.40) between the limits $\theta = 0$ and $\theta = 2\pi$, the rotation $\Delta\omega$ of the line of apsides during one orbital period of the satellite becomes

$$\Delta\omega \;=\; \frac{3\pi J_2 R_e^2}{2p^2}\,(5 \cos^2 i - 1) \tag{7.9}$$

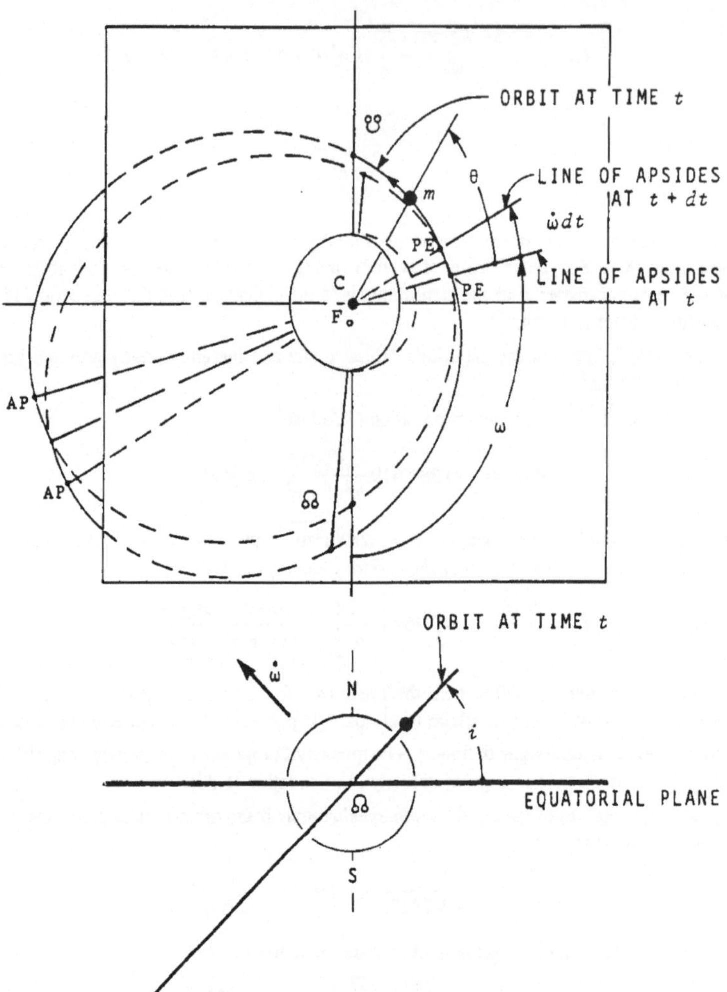

FIGURE 7.3. Rotation of the line of apsides.

Using the Kepler period $\tau = 2\pi\sqrt{a^3/\mu}$ and the constants for the Earth as the master, the *secular rotation of the line of apsides* becomes

$$\bar{\dot{\omega}} = \frac{\Delta\omega}{\tau} = 4.982 \left[\frac{R_e}{a}\right]^{7/2} \frac{5\cos^2 i - 1}{(1-\varepsilon^2)^2} \frac{\text{degree}}{\text{day}} \tag{7.10}$$

Equation (7.10) represents an *average* velocity. From it, we can conclude that a satellite near the Earth $(a \approx R_e)$ on a near equatorial orbit $(i \approx 0°)$, of near zero ellipticity $(\varepsilon \approx 0)$, has a forward rotation of the line of apsides of almost 20° during one day.

The secular rotation of the line of apsides vanishes for satellites whose orbits exhibit

$$\cos i = \frac{1}{5} \tag{7.11}$$

i.e. have a *critical* orbit inclination of

$$i = 63.435°$$

or

$$i = 116.565°$$

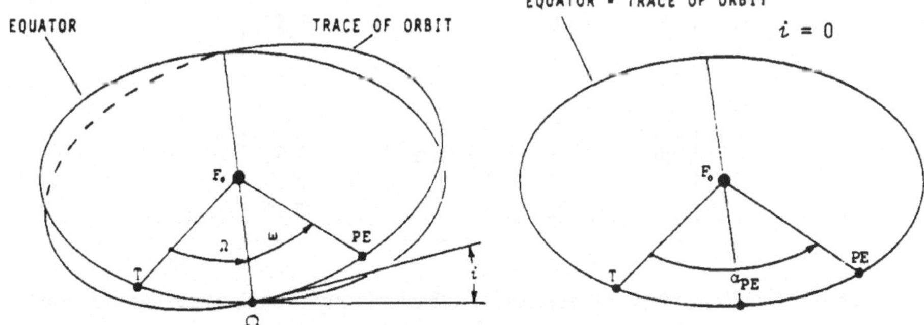

FIGURE 7.4. The right ascension of the periapsis for equatorial orbits.

The rotation of the line of apsides is retrograde for inclinations between 63.435° and 116.565°.

For equatorial satellites ($i = 0°$), the line of apsides rotates according to

$$\bar{\dot{\omega}} = \frac{6\pi J_2}{\tau} \frac{R_e^2}{a^2(1-\varepsilon^2)^2} \tag{7.12}$$

The line of nodes (which for equatorial satellites, is no longer defined) would process according to

$$\bar{\dot{\Omega}} = -\frac{3\pi J_2}{\tau} \frac{R_e^2}{a^2(1-\varepsilon^2)^2} \tag{7.13}$$

Right ascension Ω and periapsis argument ω lie now in one plane. Therefore, a new angle (Figure 7.4)

$$\alpha_{PE} = \Omega + \omega \tag{7.14}$$

can be defined, which might be referred to as the *right ascension of periapsis*. The average secular rotation of the right ascension of periapsis amounts to

$$\bar{\dot{\alpha}}_{PE} = \frac{3\pi J_2}{\tau} \frac{R_e^2}{a^2(1-\varepsilon^2)^2} \tag{7.15}$$

It is only applicable for satellites in equatorial orbits.

7.3 Major Axis, Eccentricity, Inclination

If one enters the components of the perturbation force, as given by equations (5.46), (5.47) and (5.48) into equations (6.17), (6.25) and (6.43) and integrates the resulting expressions over one full orbit, i.e. between $\theta = 0$ and $\theta = 2\pi$, one finds that the secular variations of the semi-major axis, eccentricity, and inclination all add up to zero, i.e.

$$\Delta a = 0 \tag{7.16}$$

$$\Delta \varepsilon = 0 \tag{7.17}$$

$$\Delta i = 0 \tag{7.18}$$

See also Table 7.1.

$$\Delta a \;\; = \;\; 0 \tag{7.16}$$

$$\Delta \varepsilon \;\; = \;\; 0 \tag{7.17}$$

$$\Delta \Omega \;\; = \;\; -3\pi J_2 \frac{R_e^2}{p^2} \cos i \tag{7.5}$$

$$\Delta \omega \;\; = \;\; \frac{3}{2}\pi J_2 \frac{R_e^2}{p^2} (5\cos^2 i - 1) \tag{7.9}$$

$$\Delta i \;\; = \;\; 0 \tag{7.18}$$

TABLE 7.1. Secular variations of the geometric orbital elements of an Earth satellite, per orbit.

Problems

7.1 Given is the Kepler ellipse of a satellite. A small constant transverse perturbation force f_θ is acting on the satellite. (a) Determine the change Δa of the semi-major axis for one orbit; (b) compute Δa for a satellite of 500 kg mass orbiting the Earth, if a = 10 000 km, ε = 0.001, f_θ = 0.2 N; and (c) computer the ratio f_θ / K, where K is the Kepler force.

7.2 The precession of the orbital plane for one orbit around the Earth is given by

$$\Delta \Omega \;\; = \;\; -3\pi J_2 \frac{R_e^2}{p^2} \cos i$$

The satellite "Explorer 36" ($\uparrow$ 1968-January-11) had a Kepler period of 112.2 min and an orbit inclination of 105.8°. The perigee was at 1080 km above the Earth (sphere). By how much did the right ascension of the ascending node change in one day?

7.3 Given is the Kepler ellipse of a satellite. A small constant radial perturbation force f_r acts on the satellite. Determine the change Δa of the semi-major axis a for one orbit. Discuss the result for the case of a circular orbit.

7.4 Satellite "ESRO 2" ($\uparrow$ 1968-May-17) had an elliptic orbit, with a perigee of 350 km and an apogee of 1100 km about the surface of the Earth (sphere). The orbit was polar. By how much did the perigee move during the one orbit?

7.5 What was the argument of perigee of the satellite "ISIS II" ($\uparrow$ 1971-March-31) on 1971-May-26, if at the same time of day on 1971-May-26, it was 140.7°? The perigee of the orbit was at 1354 km, the apogee at 1421 km height about the surface of the Earth (sphere). The orbit inclination was 88.153°.

7.6 The satellite "SPOT I" ($\uparrow$ 1986-February-21) has a circular orbit whose plane has an inclination of 98.37° and is sun-synchronous. What is the altitude, above the Earth sphere, of the satellite orbit?

8

Orbital Periods

The spheroidal shape of the Earth causes a deviation of the *absolute period*, i.e. the time required to pass through 360°, from the Kepler period of an Earth satellite.

In addition, it is worthwhile to determine other periods. We have already found that the ascending node carries out a retrograde motion. The orbital angle from passage through ascending node to passage through the subsequent ascending node is thus smaller than 360° and one can consequently define a *nodal period*.

The slow progression of the perigee affects the time between two successive passages of a satellite through the perigee (or periapsis) of its orbit, leading to the definition of an *apsidal period*.

8.1   Absolute Period

Since it is known from previous work, i.e. from equations (7.16) and (7.17), that the secular variations of the semi-minor axis and of the eccentricity of a point satellite in a gravitational potential (7.1) of

$$U = -\frac{\mu m}{r}\left[1 - \frac{1}{2}J_2\,\frac{R_e^2}{r^2}\,(3\sin^2\delta - 1)\right]$$

vanish, the derivation of the absolute period can be based on the assumption, that the semi-major axis and the eccentricity of the orbit remain constant. During its orbit, the satellite is, however, not only attracted by a Kepler force (1.19) of

$$\mathbf{K} = -\frac{\mu m}{r^2}\mathbf{e}_r$$

but, in addition, also by the perturbation force **f**, whose radial component is given by equation (5.47). Adding the two, the total radial force of attraction is

$$\mathbf{F}_r = -\frac{\mu m}{r^2}\left[1 - \frac{3}{2}J_2\,\frac{R_e^2}{r^2}\,(3\sin^2 i\,\sin^2 u - 1)\right]\mathbf{e}_r \tag{8.1}$$

We further assume that h = constant is still applicable, but we replace the attraction constant μ in the equation for the angular momentum h by

$$\mu\left[1 - \frac{3}{2}J_2\,\frac{R_e^2}{r^2}\,(3\sin^2 i\,\sin^2 u - 1)\right] \tag{8.2}$$

In doing this, we obtain

$$h = r^2\dot\theta = \sqrt{\mu p\left[1 - \frac{3}{2}J_2\,\frac{R_e^2}{r^2}\,(3\sin^2 i\,\sin^2 u - 1)\right]} \tag{8.3}$$

leading to

$$dt = \frac{r^2 d\theta}{\sqrt{\mu p\left[1 - \frac{3}{2}J_2\,\frac{R_e^2}{r^2}\,(3\sin^2 i\,\sin^2 u - 1)\right]}} \tag{8.4}$$

The second term under the root sign is so small that one can employ a series expansion, which may be broken off after the second term,

$$dt = \frac{r^2}{\sqrt{\mu p}}\left[1 + \frac{3}{4}J_2\,\frac{R_e^2}{r^2}\,(3\sin^2 i\,\sin^2 u - 1)\right]d\theta \tag{8.5}$$

Let us integrate the left side between the limits 0 and $\tau_{2\pi}$, and the right side between 0 and 2π. The first term on the right side is the well-known Kepler period τ, such that

$$\tau_{2\pi} = \tau + \frac{3}{4}J_2 \frac{R_e^2}{\sqrt{\mu p}} \int_0^{2\pi} (3\sin^2 i \, \sin^2 u - 1)d\theta \tag{8.6}$$

Considering that $u = \theta + \omega$ and approximating $ab \approx p^2$, we obtain for the *absolute period*

$$\tau_{2\pi} = \tau\left[1 - \frac{3}{8}J_2 \frac{R_e^2}{p^2}(3\cos^2 i - 1)\right] \tag{8.7}$$

If one expresses equation (8.7) as

$$\tau_{2\pi} = \tau + \Delta\tau_{2\pi}$$

then one obtains for the variation $\Delta\tau_{2\pi}$ of the absolute period for one orbit

$$\Delta\tau_{2\pi} = -\frac{3}{8}\tau J_2 \frac{R_e^2}{p^2}(3\cos^2 i - 1) \tag{8.9}$$

or

$$\Delta\tau_{2\pi} = -\frac{3}{4}\pi\sqrt{\frac{a^3}{\mu}} J_2 \frac{R_e^2}{p^2}(3\cos^2 i - 1) \tag{8.10}$$

Equation (8.10) is listed in Table 8.1, equation (8.7) in Table 8.2 for easy reference.

As an example, select a typical near-Earth satellite, on an equatorial orbit, with $a = 7000$ km and $\varepsilon = 0.05$. We also have $R_e = 6378.16$ km, $J_2 = 1082.628 \cdot 10^{-6}$, $\mu = 398\,601.19$ km^2/s^3, $p = a(1 - \varepsilon^2) = 6\,982.5$ km, and $i = 0°$.

The Kepler period for this satellite is

$$\tau = 2\pi\sqrt{\frac{a^3}{\mu}} = 5828.5\,\text{s} = 97.14\,\text{min}$$

The variation $\Delta\tau_{2\pi}$ is

$$\Delta\tau_{2\pi} = -\frac{3}{8} \cdot 5828.5 \cdot 1082.628 \cdot 10^{-6}\left[\frac{6378.16}{6982.5}\right]^2 \cdot 2 = -3.949\,\text{s}$$

Thus a satellite near the Earth, on an equatorial orbit, has an absolute period, which is some four seconds *shorter* than its Kepler period. On a polar orbit, the absolute period is some two seconds *longer* than the Kepler period, because with $i = 90°$, one gets

$$\Delta\tau_{2\pi} = \frac{3}{8} 5828.5 \cdot 1082.628 \cdot 10^{-6} \cdot \left[\frac{6378.16}{6982.5}\right]^2 = 1.974\,\text{s}$$

As can be deduced from equation (8.10), there is no difference between absolute period and Kepler period in case a satellite moves on an orbit, whose plane has an inclination i for which

$$3\cos^2 i - 1 = 0 \tag{8.11}$$

i.e. for orbital plane inclinations of

$$i = 54.736° \tag{8.12a}$$

and

$$i = 125.264° \tag{8.12b}$$

8.2 Nodal Period

Measured with respect to an inertial system, the ascending node of a satellite orbit moves by an angle of $\Delta\Omega$ per orbit. For the position vector of a satellite, this means that, after passage through the ascending node and upon sweeping out 360° (with respect to the inertial system), the succeeding ascending node, which has moved on in the meantime, has not yet been reached (Figure 8.1). In order to reach it, the satellite's position vector has to sweep out an additional angle of magnitude

$$\Delta\Omega \cos i \tag{8.13}$$

Altogether the satellite needs a time period

$$\tau_\Omega \;=\; \frac{2\pi + \Delta\Omega \cos i}{2\pi}\tau_{2\pi} \tag{8.14}$$

in order to move from ascending node to successive ascending node. With the aid of equation (7.5), the nodal period can then be expressed as

$$\tau_\Omega \;=\; \tau_{2\pi}\left[1 - \frac{3}{2}J_2\,\frac{R_e^2}{r^2}\cos^2 i\right] \tag{8.15}$$

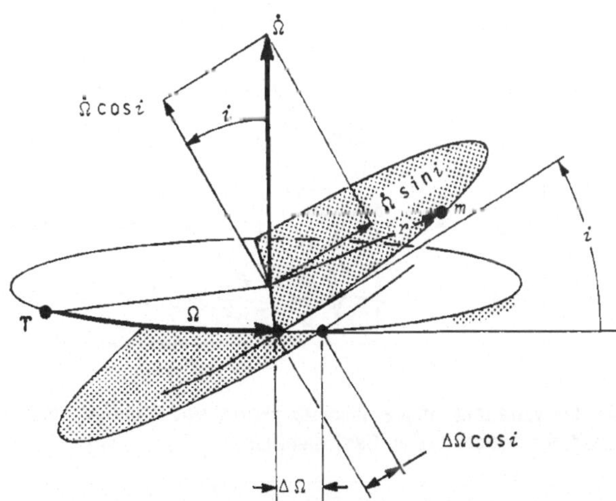

FIGURE 8.1. Progression of the ascending node during one orbit.

Introducing the Kepler period, and upon neglect of a very small term containing J_2^2, the *nodal period* is eventually obtained as

$$\tau_\Omega \;=\; \tau\left[1 - \frac{3}{8}J_2\,\frac{R_e^2}{p^2}(7\cos^2 i - 1)\right] \tag{8.16}$$

or

$$\tau_\Omega \;=\; 2\pi\sqrt{\frac{a^3}{\mu}}\left[1 - \frac{3J_2 R_e^2(7\cos^2 i - 1)}{8a^2(1-\varepsilon^2)^2}\right] \tag{8.17}$$

See also Table 8.2.

The difference between nodal period and Kepler period is

$$\Delta\tau_\Omega \;=\; \tau_\Omega - \tau \;=\; -\frac{3}{3}\tau J_2\,\frac{R_e^2}{p^2}(7\cos^2 i - 1) \tag{8.18}$$

See also Table 8.1.

As an example, we look at a near-Earth satellite again, on an equatorial orbit. We obtain

$$\Delta\tau_\Omega = -\tau \frac{3J_2}{8}(7-1) = -11.844\,\text{s}$$

i.e. the nodal period is some 12 s *shorter* than the Kepler period for a satellite on an equatorial orbit. For the same satellite, but on a polar orbit, we obtain

$$\Delta\tau_\Omega = -\tau \frac{3J_2}{8}(0-1) = +1.974\,\text{s}$$

i.e. some two seconds longer than the Kepler period.

Kepler period τ and nodal period τ_Ω of an Earth satellite are the same, if in equation (8.16),

$$7\cos^2 i - 1 = 0 \tag{8.19}$$

This condition is satisfied for orbital inclinations of

$$i = 67.792° \tag{8.20a}$$

and

$$i = 112.208° \tag{8.20b}$$

$$\boxed{\begin{aligned}
\Delta\tau_{2\pi} &= -\frac{3}{4}\pi\sqrt{\frac{a^3}{\mu}}\,J_2\frac{R_e^2}{p^2}(3\cos^2 i - 1) &\tag{8.10}\\[2ex]
\Delta\tau_\Omega &= \frac{3}{4}\pi\sqrt{\frac{a^3}{\mu}}\,J_2\frac{R_e^2}{p^2}(7\cos^2 i - 1) &\tag{8.18}\\[2ex]
\Delta\tau_{PE} &= \frac{3}{4}\pi\sqrt{\frac{a^3}{\mu}}\,J_2\frac{R_e^2}{p^2}(3\cos^2 i - 1) &\tag{8.24}
\end{aligned}}$$

TABLE 8.1. Secular variations of the absolute period, nodal period, and apsidal period from Kepler period, for one orbit of an Earth satellite.

8.3 Apsidal Period

If a satellite has passed the periapsis of its orbit and its position vector has subsequently swept out an angle of 360° (with respect to an inertial reference system), a time interval of $\tau_{2\pi}$ has gone by. The satellite, however, has not regained the periapsis of its orbit, because the periapsis has meanwhile moved through an angle of $\Delta\omega$ as measured from the nodal line. The nodal line itself has moved on, and that by an amount of $\Delta\Omega'$ whose projection onto the orbital plane is $\Delta\Omega\cos i$ (Figure 8.1). This means that the satellite has to traverse an angle of $\Delta\omega + \Delta\Omega\cos i$, in addition to 360°, to regain periapsis. The total time spent between two periapsis passages is thus

$$\tau_{PE} = \frac{2\pi + \Delta\omega + \Delta\Omega\cos i}{2\pi}\tau_{2\pi} \tag{8.21}$$

We make use of equations (7.5) and (7.9), and obtain

$$\tau_{PE} = \left[1 + \frac{3}{4}J_2\frac{R_e^2}{p^2}(3\cos^2 i - 1)\right]\tau_{2\pi} \tag{8.22}$$

After introducing the Kepler period from equation (8.7) and upon neglecting a very small term containing the factor J_2^2, we arrive at an equation for the *apsidal period* given by

$$\tau_{PE} = \tau\left[1 + \frac{3}{8}J_2\frac{R_e^2}{p^2}(3\cos^2 i - 1)\right]$$
(8.23)

See also Table 8.2.

For the difference $\Delta\tau_{PE}$ between apsidal period and Kepler period, one obtains

$$\Delta\tau_{PE} = \tau_{PE} - \tau = \frac{3}{8}\tau J_2\frac{R_e^2}{p^2}(3\cos^2 i - 1)$$
(8.24)

See also Table 8.1.

Using our example of a near-Earth satellite again, we can determine the apsidal period on an equatorial orbit, and find it to be some four seconds greater than the Kepler period. The same satellite on a polar orbit has an apsidal period, which is some two seconds shorter than its Kepler period.

The difference between apsidal period and Kepler period vanishes when $3\cos^2 i - 1 = 0$, i.e. for orbit inclinations of

$$i = 54.736°$$
(8.25a)

and

$$i = 125.267°$$
(8.25b)

$$\tau = 2\pi\sqrt{\frac{a^3}{\mu}}$$

$$\tau_{2\pi} = \tau\left[1 - \frac{3}{8}J_2\frac{R_e^2}{p^2}(3\cos^2 i - 1)\right] \quad (8.7)$$

$$\tau_\Omega = \tau\left[1 - \frac{3}{8}J_2\frac{R_e^2}{p^2}(7\cos^2 i - 1)\right] \quad (8.16)$$

$$\tau_{PE} = \tau\left[1 + \frac{3}{8}J_2\frac{R_e^2}{p^2}(3\cos^2 i - 1)\right] \quad (8.23)$$

TABLE 8.2. Kepler period, absolute period, nodal period, and apsidal period.

Problems

8.1   A satellite of the Earth has an orbital inclination of 90°, an eccentricity of 0.2, and a perigee 7000 km from Earth centre. Compute the satellite's (a) Kepler period; (b) absolute period; (c) nodal period; and (d) apsidal period.

8.2   Satellite "Azur" ($\uparrow$ 1969-November-08) had a perigee of 383 km and an apogee of 3145 km height above the Earth (sphere). Its orbital plane had an inclination of 102.975°. Compute the satellite's (a) Kepler period; (b) absolute period; (c) nodal period; and (d) apsidal period.

8.3   A satellite of the Earth has an orbital inclination of 67.792°, an eccentricity of 0.1, and a perigee of 6700 km from Earth centre. Computer the satellite's (a) Kepler period; (b) absolute period; (c) nodal period; and (d) apsidal period.

8.4   If the simplification $ab \approx p^2$ is not made, what would equation (8.7) look like? Calculate the orbital period (a) according to the new formula, and (b) according to equation (8.7), for an Earth satellite on an orbit with p = 7500 km, ε = 0.1, i = 30°.

8.5   A Jupiter satellite on a near-polar orbit has an orbit inclination of 86° and an eccentricity of 0.1. Its perigee is at 72 818 km from Jupiter centre. Compute the satellite's (a) Kepler period; (b) absolute period; (c) nodal period; and (d) apsidal period.

9

Other Perturbations

In addition to the gravitational force, there are several other, much smaller, forces that are encountered by a satellite. They lead to perturbations of the satellite's orbit. Such forces are caused by:

1. Air resistance (neutral particle drag)

2. Solar wind (maximum pressure near Earth $p = 9.02 \, \mu\text{Pa}$)

3. Meteorite impact

4. Magnetic field

5. Electrical charge of the ionosphere (charged particle drag)

6. Thrust by satellite-borne rockets

The effects of each of these influences on a satellite's orbit are similar. Each is represented by a small force acting on the satellite. For satellites whose orbits are close to Earth, the air resistance of the Earth's atmosphere represents usually the greatest single effect amongst those listed, and we shall therefore discuss its influence in some detail. This is followed by a short description of the solar wind.

Other problems encountered by satellites are due to temperature differences within the satellite brought about by solar radiation. They include shape distortion due to thermal bending, and oscillation due to thermal flutter.

Meteorite impact, magnetic field, and charged particle drag effects are not dealt with in the present book. Thrust effects were discussed in Chapter 4.

9.1 Air Resistance

A detailed description of the atmosphere conditions in the uppermost reaches of the air layer surrounding the Earth is very difficult because of continuously and rapidly changing conditions.

The density of the air becomes smaller and smaller, the greater the height above the Earth's surface. The so-called *upper atmosphere* is the region beginning at 80 km height above the Earth's surface. Conditions in the upper atmosphere differ markedly from those in the lower atmosphere. The density in the upper atmosphere in particular, is subject to large variations, depending on locality and time.

The main influences of the air resistance upon a satellite orbit can fortunately be derived without having to have a detailed knowledge of the density variations. To this end, we shall make certain simplifying assumptions.

1. The satellite is, on the one hand, so small that it can be looked upon as a point satellite moving in the gravitational field of its master.

2. On the other hand, the satellite is large enough to be subject to air resistance.

3. The satellite has spherical symmetry, in the sense that air resistance does not cause a torque about the satellite's mass centre.

4. The Earth's atmosphere has spherical symmetry.

5. The Earth's atmosphere is at rest with respect to outer space, i.e. it does not participate in the Earth's spin, nor is there any wind.

On the basis of the assumptions listed, we conclude that a perturbation force

$$\mathbf{f} = -\frac{c_d}{2} A \, \rho v^2 \, \mathbf{e}_t \tag{9.1}$$

acting on the satellite will represent the drag due to air resistance. The quantity ρ is the density of the air, v is the speed and A is the frontal area of the satellite, c_d is the air *drag* (or *resistance*) *coefficient*, which has values between 2.0 and 2.6 for typical satellites at heights from 150 to 400 km above the Earth's surface, and e_t is the unit vector in orbit tangential direction (Figure 9.1). The air resistance coefficient c_d rises to about 4.0 at 600 km height, and to larger values beyond.

At about 600 km height, electrically charged particle drag, solar wind effects, and magnetic field drag begin to become more dominant than air resistance. Altogether, though, the perturbation forces at more than 600 km height are so insignificant that they usually pose no problem.

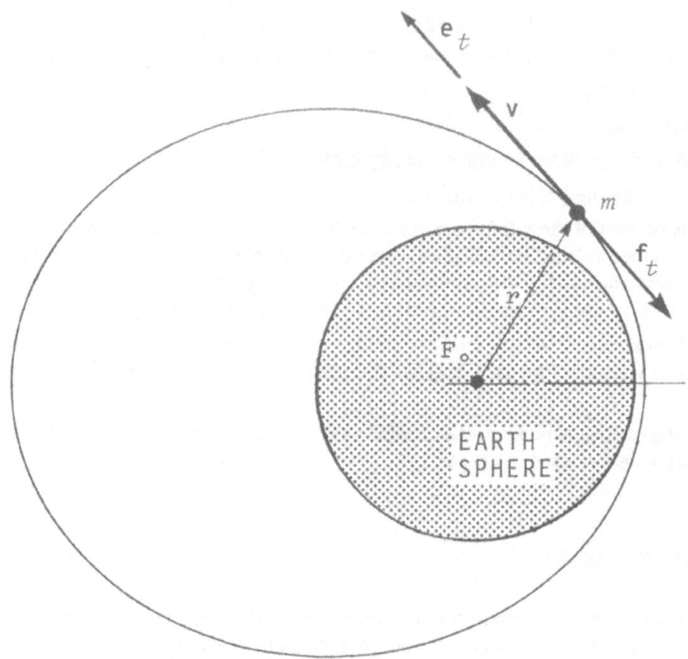

FIGURE 9.1. Air resistance f_t.

9.2 Circular Orbit

For a circular orbit, tangential and transverse direction coincide. We shall now compute the influence of the air resistance upon the various geometrical orbit elements by making use of the equations for the variation of elements. We assume that the air resistance remains constant throughout each orbit.

The effect upon the semi-major axis is obtained by using equation (6.17), with $f_r = 0$, $f_\theta = f_t$, and $\varepsilon = 0$,

$$\frac{da}{d\theta} = \frac{2p^3}{\mu m} f_t \tag{9.2}$$

Integrated over one orbit with the limits 0 and Δa, and 0 and 2π, one obtains

$$\Delta a = \frac{4\pi p^3}{\mu m} f_t \tag{9.3}$$

or, with the aid of equation (9.1),

$$\Delta a = -\frac{2\pi c_d A \rho v^2 p^3}{\mu m} \tag{9.4}$$

For circular orbits, one has $p = a$ and $v^2 = \mu/a$ such that

$$\Delta a = -\frac{2\pi c_d A \rho a^2}{m} \qquad (9.5)$$

The effect of the air resistance upon the eccentricity is obtained from equation (6.25) and amounts to

$$\Delta \varepsilon = \frac{2p^2 f_t}{\mu m} \int_0^{2\pi} \cos\theta = 0 \qquad (9.6)$$

An investigation of the effects of the air resistance upon the ascending node, equation (6.31), and upon the orbit inclination, equation (6.42), shows that both are zero as well.

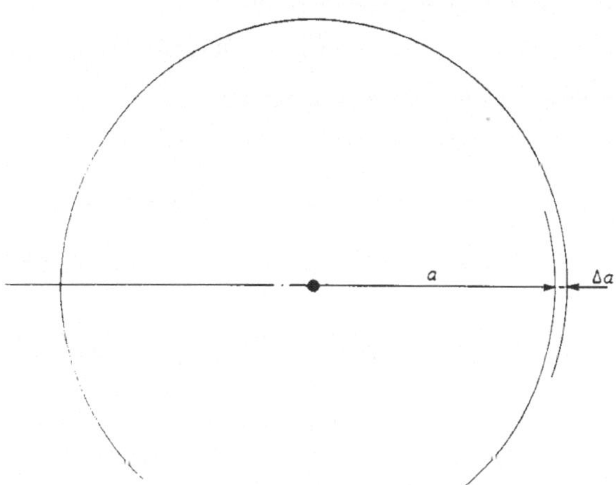

FIGURE 9.2. A small change in orbit size in case of an essentially circular orbit.

The orbital period is reduced as the orbit radius becomes smaller (Figure 9.2). The first partial derivative of the Kepler period (2.3) with respect to the orbital semi-major axis is

$$\frac{\partial \tau}{\partial a} = 3\pi\sqrt{\frac{a}{\mu}} \qquad (9.7)$$

If this equation is used for the change in orbital time, one gets

$$\Delta \tau = 3\pi\sqrt{\frac{a}{\mu}}\,\Delta a \qquad (9.8)$$

or, with the help of equation (9.5),

$$\Delta \tau = -\frac{3\pi c_d A \rho a \tau}{m} \qquad (9.9)$$

When a satellite is subjected to air resistance, i.e. subject to a "braking" effect, the satellite's speed actually increases. This apparent paradox can be explained by the fact that the reduction in potential energy is twice as large as the gain in kinetic energy of the satellite, resulting in an overall energy decrease in spite of a kinetic energy increase.

The velocity increase per orbit is obtained by using the derivative of the speed equation (2.9)

$$2v\,dv = -\frac{\mu}{a^2}\,da \qquad (9.10)$$

which, by using equation (9.5) and by introducing Δv instead of dv, becomes

$$\Delta v = \frac{\pi c_d A \rho a v}{m} \qquad (9.11)$$

The total energy of the satellite is, according to equation (2.35),

$$T \; + \; U \; = \; -\frac{\mu m}{2a} \tag{9.12}$$

The change of total energy can be obtained by differentiating equation (9.12),

$$d(T + U) \; = \; \frac{\mu m}{2a^2}\, da \tag{9.13}$$

a relationship which can be used to obtain an expression for the energy change per orbit

$$\Delta(T + U) \; = \; \frac{\mu m}{2a^2}\, \Delta a \tag{9.14}$$

With the aid of equation (9.5), the energy change becomes

$$\Delta(T + U) \; = \; -\pi\mu c_d A \rho \tag{9.15}$$

It is negative, thus it represents a *loss* of energy, as expected.

The potential energy portion alone, equation (2.29), has changed by

$$\Delta U \; = \; \frac{\mu m}{a^2}\Delta a \; = \; -2\pi\mu c_d A \rho \tag{9.16}$$

Since $\Delta v \; = \; -\mu\Delta a/2a^2 v$ according to equation (9.10), we find that the kinetic energy alone has increased by

$$\Delta T \; = \; mv\Delta v \; = \; -\frac{\mu m}{2a^2}\Delta a \; = \; \pi\mu c_d A \rho \tag{9.17}$$

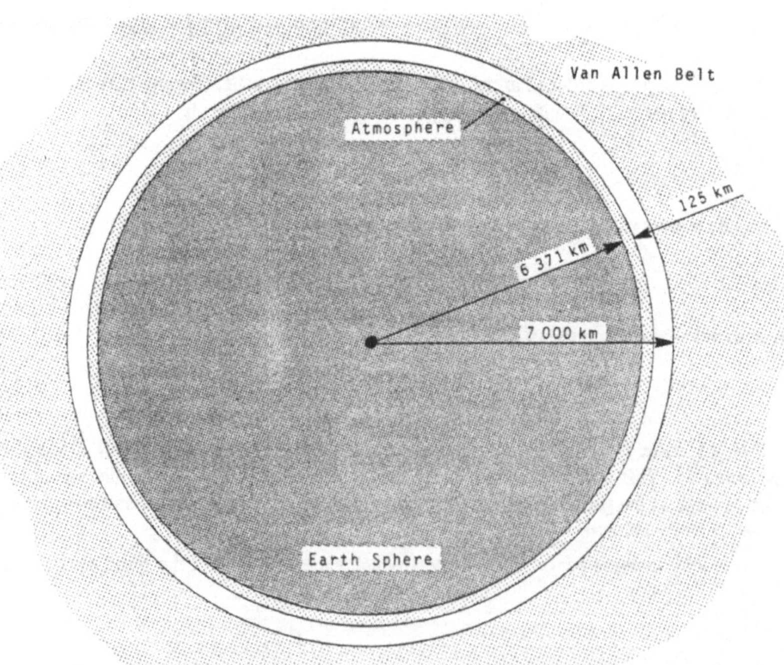

FIGURE 9.3. Manned space missions over extended time periods are possible in a narrow belt between atmosphere and Van Allen Belt, and above the Van Allen Belt.

9.3   Upper Limit of the Atmosphere

The assumption that an orbit is circular is only permitted as long as the changes in orbital elements are small enought to justify an approximation of the real orbit (of spiral shape) by a circular orbit.

The satellite will descend steeply when the air resistance is so great that the orbit assumes a pronounced spirality. The height above Earth surface, at which this event commences, depends also on the size and shape of the frontal area of the satellite. The limit, below which a circular satellite orbit can generally no longer be maintained, is at about 125 km above the Earth's surface (Figures 9.3 and 9.4). For *manned* space missions, there is also an upper limit at about 600 km caused by exposure to unacceptable irradiation levels in the Van Allen Radiation Belt[*]. The Van Allen Radiation Belt reaches to about 9000 km above Earth, beyond which extended time manned space missions are again possible (Minzner, 1976; Wildmann, 1980; Lanzerotti, 1981; Jacchia, 1977).

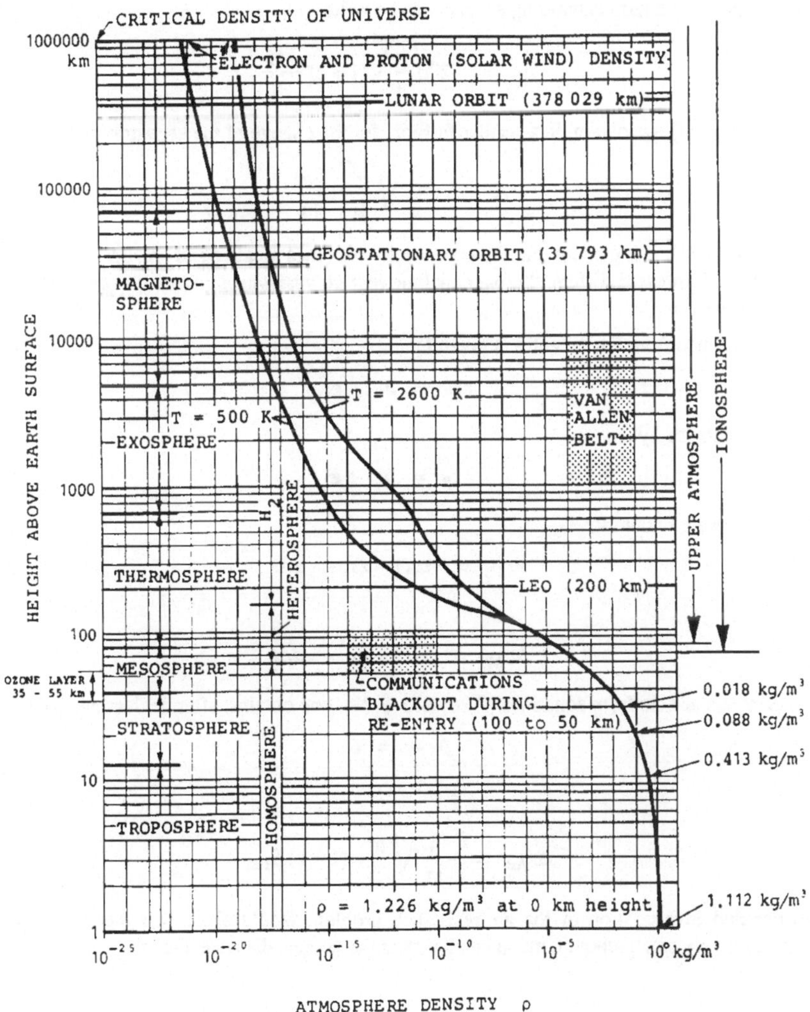

FIGURE 9.4. Atmospheric density within Earth's influence sphere.

---

[*] James Alfred Van Allen (1914-   ), American physicist.

9.4   Elliptic Orbit

A satellite on a elliptical orbit reaches its maximum velocity in the perigee. As the perigee is also the low point of the orbit, it is the point with the highest air density. Adding the two characteristics permits an approximate treatment, by assuming that there is just *one* impulse imparted to the satellite, and that it occurs while passing through the perigee. We shall refer to this impulse as

$$\mathbf{S} = -S_{PE}\,\mathbf{e_\theta} \tag{9.18}$$

From equation (6.14) with $f_r = 0$ and $f_\theta = -S_{PE}/\Delta t$ at $\theta = 0°$, one obtains for the change of semi-major axis per orbit

$$\Delta a = -\sqrt{\frac{p}{\mu}}\,\frac{2p}{(1-\varepsilon^2)^2}\,(1+\varepsilon)\,\frac{S_{PE}}{m} \tag{9.19}$$

which, after neglecting a term containing ε^2, can also be written

$$\Delta a = -2\sqrt{\frac{a^3}{\mu}}\,(1+\varepsilon)\,\frac{S_{PE}}{m} \tag{9.20}$$

From equation (6.24), we obtain, in a similar fashion, for the change of the eccentricity per orbit

$$\Delta\varepsilon = -\frac{2}{a}\sqrt{\frac{a^3}{\mu}}\,\frac{S_{PE}}{m} \tag{9.21}$$

From equation (9.21), it is evident that the eccentricity becomes smaller and smaller as a consequence of the air resistance. Expressed differently, we conclude that *air resistance causes the orbit shape to become more and more circular*.

The perigee distance from Earth centre is given by

$$r_{PE} = a(1-\varepsilon) \tag{9.22}$$

and the apogee distance by

$$r_{AP} = a(1+\varepsilon) \tag{9.23}$$

The changes per orbit amount to

$$\Delta r_{PE} = \Delta a(1-\varepsilon) - a\Delta\varepsilon \tag{9.24}$$

and

$$\Delta r_{AP} = \Delta a(1+\varepsilon) + a\Delta\varepsilon \tag{9.25}$$

After entering Δa and $\Delta\varepsilon$ from equations (9.20) and (9.21), one obtains, after neglecting terms with ε^2,

$$\Delta r_{PE} = 0 \tag{9.26}$$

and

$$\Delta r_{AP} = -\frac{4}{m}\sqrt{\frac{a^3}{\mu}}\,(1+\varepsilon)\,S_{PE} \tag{9.27}$$

A more detailed investigation of the air resistance problem would show that the perigee distance is reduced too, by a very small amount, much smaller than the apogee distance change.

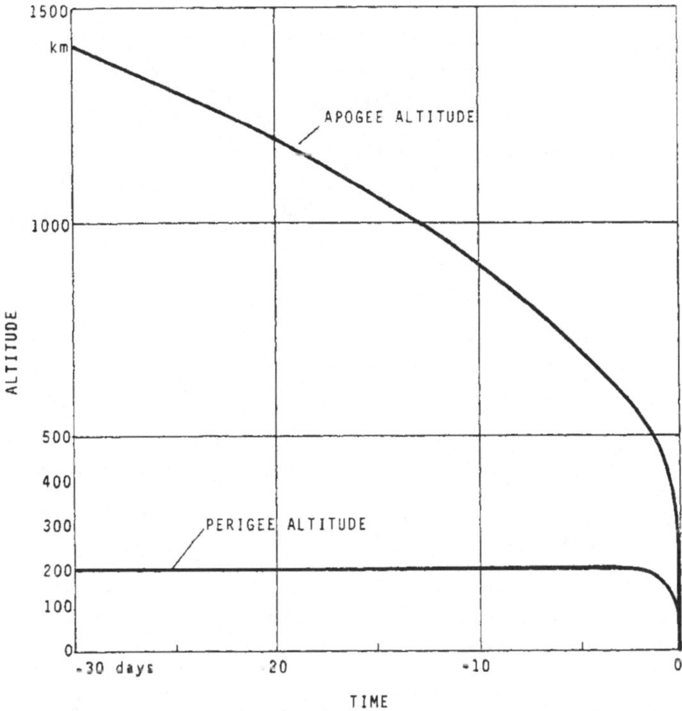

FIGURE 9.5. Apogee and perigee altitude of Atlas Score satellite ($\downarrow$ 1959-January-21).

We have based our derivation on the assumption that the magnitude S_{PE} of the impulse in the perigee is defined by the satellite speed and the air density, both in the perigee. How are these two affected by the changes in the orbit shape brought about by air resistance? From equation (9.26), we learn that the perigee height remains constant, and we conclude that consequently the air density remains constant as well. In order to determine the change of satellite speed in the perigee, we make use of the speed equation (2.13) which eventually provides us with

$$\Delta v_{PE} = -\frac{S_{PE}}{m} \tag{9.28}$$

When the aid of the speed equation (2.9), we can also determine the change of the speed of passage through the apogee,

$$\Delta v_{AP} = 3\frac{S_{PE}}{m} \tag{9.29}$$

A comparison of equations (9.28) and (9.29) indicates that while the satellite's speed of passage through the perigee of its orbit is reduced, the speed of passage through its apogee is increased, by an amount three times as high.

Since it is awkward (Holbrook, 1975) to compute the impulse magnitude S_{PE} from velocity and density data, we shall endeavour to obtain an idea of a typical value for S_{PE} by considering a satellite of 200 kg mass, with a perigee of 140 km above Earth surface, and an apogee of 1000 km above Earth surface. If the apogee height is reduced to 999.5 km after one orbit, what is the magnitude of the impulse in the perigee?

We calculate the eccentricity of the orbit

$$\varepsilon = \frac{r_{AP} - r_{PE}}{r_{AP} + r_{PE}} = 0.06195$$

and the semi-major axis

$$a \;=\; \frac{1}{2}(r_{AP} + r_{PE}) \;=\; 60414 \,\text{km}$$

and then solve equation (9.27) for the impulse

$$S_{PE} \;=\; -\frac{\sqrt{\mu}\, m}{4\sqrt{a^3}\,(1+\varepsilon)}\, \Delta r_{AP}$$

giving

$$S_{PE} \;=\; 25.7 \,\text{N s}$$

9.5   The Solar Wind

The Earth is constantly exposed to a wind coming from the Sun. This *solar wind* hits our planet at about nine times the speed of sound, that is, at Mach 9. This causes the Earth to produce a shockwave in this current.

The first indications pointing to a solar wind were discovered while observing the tail of comets. Various photographs of comets showed that the tail trailed by the spherical head of the comet was split up, and that one part of this split tail turned away from the Sun. It looked just as though the Sun blew against this part. Meantime, several satellites and space-probes following very eccentric orbits around the Earth, such as Lunik II and III, Explorer 10, Mariner 2, and above all the Interplanetary Measuring Platform (IMP) of NASA, have penetrated far into the solar wind and have measured it directly. The measured values showed wind velocities ranging from 300 to 800 km/s and a temperature of nearly 100 000°C. This wind from the Sun chiefly takes the form of hydrogen gas in an ionized form, that is, a mixture of electrons and protons, or nuclei of hydrogen atoms respectively, the so-called gas plasma. The density of the wind-matter measured so far amounts to no more than a few protons per cubic centimetre, and is thus less than that in a good terrestrial vacuum. But due to the high speed of the solar wind, about 100 million protons pass through an area of one square centimetre in one second in the vicinity of Earth.

The shock-wave caused by the current's meeting with Earth, naturally does not form upon direct impact of the solar wind on Earth's atmosphere or its actual surface - it never gets as far as this - but at a distance of several Earth radii, where the protons and electrons of the solar wind encounter the geo-magnetic field. Being electrically charged particles, they cannot simply penetrate a magnetic field or break out of a magnetic field. This has been confirmed by measurements, which have also served to prove the existence of the Earth's shock-wave. It is not yet quite clear how far the solar wind can move on a straight line on its way from the Sun into space, or the extent to which it may be forced to make detours - say, on a spiral track - by magnetic fields. It may also well be that the solar wind forms eddies, which in turn form a magnetic field - like the electric current in a wire coil - and carry this along with them.

At heights of 180 km above the Earth's surface, i.e. still deep within Earth's magnetic field, solar wind velocities of 55-70 km/s have been measured, causing a pressure of about 9.02 µPa on a fully reflecting surface normal to the solar wind, or 4.51 µPa on a black body.

If we take the trouble to compute the amount of matter constantly lost by the Sun in emitting the solar wind, we arrive at a figure of about *one million tons* of hydrogen *per second*. It might be thought that a mass loss rate of this magnitude would be more than even a star could stand for any length of time, but a little calculation will quickly show that the mass loss over a time span of, say, one million years is insignificant indeed.

Some engineers (Forward, 1985) suggest the exploitation of the solar wind to propel spacecraft equipped with *solar sails* on deep space missions.

9.6 Satellite Temperature

Although not strictly an orbit dynamics problem, a brief digression into the question of the *temperature* of a spacecraft, or spacecraft element, as a consequence of its exposure to the Sun, may be of interest, since it leads to an understanding of thermal bending and thermal flutter, two mechanical effects which in turn have an effect on the spacecraft's motion. For a spacecraft in radiative equilibrium, the power balance is

$$\varepsilon T^4 \sigma A_E \;=\; \alpha \frac{S}{R^2} A_I \;+\; Q \tag{9.30}$$

with

ε = spacecraft surface emittance ($\varepsilon = 1$ for a black body, $\varepsilon = 0.4$ for a typical surface, $\varepsilon = 0.02$ for a silver-plated surface)

T = spacecraft surface temperature, K

$\sigma = 5.67 \cdot 10^{-8}$ W/m^{2}K^4 = Stefan-Boltzmann constant

A_E = spacecraft radiating area, m^2

α = spacecraft solar absorptance, ($\alpha = 1$ for a black body, $\alpha = 0.6$ for a typical surface, $\alpha = 0.16$ for a silver-plated surface)

S = 1353 W/m^2 = mean solar radiation constant at 1 AU

R = distance from Sun to Spacecraft, AU

A_I = spacecraft incidence area, normal to solar radiation, m^2

Q = power dissipated internally in spacecraft, W

Take as an example a silver-plated thin-walled tubular antenna of diameter d, on a spacecraft orbiting the Earth on a low orbit. With

ε = 0.02
A_E = $\pi l d$
A_I = $l d$
α = 0.16
R = 1 AU
Q = 0

the antenna temperature is obtained from

$$T^4 \;=\; \frac{\alpha \dfrac{S}{R^2} A_I}{\varepsilon \sigma A_E} \;=\; \frac{0.16 \dfrac{1353}{1}}{0.02(5.76)10^{-8}}$$

which yields

$$T \;=\; 494 \text{ K}$$

or

$$t \;=\; 221°\text{C}$$

Spacecraft appendages, such as tubular antennas, exhibit a temperature difference between the illuminated side and the dark side, that assumes large values for materials with high absorptivity, low emissivity, and low thermal conductivity. For thin-walled tubing, the temperature difference is essentially independent of the diameter, and for silver-plated beryllium-copper tubing temperature differences of 3°C have been recorded, for unplated beryllium copper tubing 10°C, and for unplated stainless steel tubing up to 40°C. Such temperature differences lead to considerable *thermal bending* (Rimrott, 1965), which in turn influence the motion of satellites (Etkin and Hughes, 1967). Another effect brought about by sunshine, that manifests itself in open section tubing, is *thermal flutter* (Rimrott, 1981).

Suggested Reading

1. Etkin, B.; P.C. Hughes. "Explanation of the Anomalous Spin Behaviour of Satellites with Long Flexible Antennas", *AIAA Journal of Spacecraft and Rockets*, 4, 9, 1967, pp. 1139-1145.

2. Forward, R.L. "Feasibility of Interstellar Travel: A Review", *Proceedings 36th International Astronautical Congress*, Stockholm, 1985, AIAA-85-489.

3. Holbrook, J.A.D. "The Applications of Model Atmospheres to Synchronous Satellite Launches", *CASI Transactions*, 8, 1, 1975, pp. 25-28.

4. Jacchia, L.G. "Thermospheric Temperature, Density, and Composition: New Models", *Smithsonian Astrophysical Laboratory*, Special Report No. 375, 1977, p. 106.

5. Lanzerotti, L.J. "Geospace - The Earth's Plasma Environment", *Astronautics and Aeronautics*, 19, 4, 1981, pp. 57-63.

6. Minzner, R.A. (Editor) "The 1976 Standard Atmosphere Above 86 km Altitude", *NASA* SP-398, 1976, pp. 1-70.

7. Rimrott, F.P.J. "Storable Tubular Extendible Member - A Unique Machine Element", *Machine Design*, 37, 28, 1965, pp. 156-165.

8. Rimrott, F.P.J. "The Frequency Criterion for Thermally Induced Vibration in Elastic Beams", *Ingenieur-Archiv*, 50, 4, 1981, pp. 281-287.

9. Wildmann, P.J.L. "Dynamics of a Rigid Body in the Space Plasma", *Space Systems and Their Interactions with Earth's Space Environment*, *AIAA*, 71, 1980, pp. 633-661.

Problems

9.1   A satellite ($m = 100$ kg) moves on a circular orbit 629 km above the surface of the Earth sphere. It has a relatively blunt ($c_d = 2.5$) frontal area of 0.9 m^2. The density of the atmosphere encountered is $1.44 \cdot 10^{-13}$ kg/m^3. By how much per orbit is the orbit radius affected?

9.2   A satellite moving about the Earth, with a perigee of 150 km above the Earth surface is affected by the atmosphere, which imparts an impulse of 30 Ns while the satellite passes through its perigee. The orbital eccentricity is 0.1. The satellite has a mass of 200 kg. Determine, for one orbit, the change of (a) semi-major axis; (b) eccentricity; (c) perigee height; (d) apogee height; (e) velocity change in perigee; and (f) velocity change for apogee.

9.3   The space shuttle ($m = 100\,000$ kg, $A = 25$ m^2, $c_d = 2$) orbits the Earth at 200 km above Earth surface. Determine the altitude loss per orbit (a) for very high density air; and (b) for very low density air.

9.4   A spacecraft on a (near) circular orbit about the Earth of 6571 km radius loses 100 m altitude/orbit due to (uniform) air drag. Compute the spacecraft's acceleration.

9.5   Measurements are to be conducted at a height of 120 km above Earth surface, where the atmospheric density is expected to be $\rho = 1.1 \cdot 10^{-7}$ kg/m^3, too high for a spaceship to operate in. Therefore, a 50 kg spherical body ($c_d = 2.1$) of 2 m diameter is to be towed on an orbit of 120 km height by an orbiter ($M = 100\,000$ kg) on a circular orbit 60 km higher. Assume zero drag at this level. Also assume that there is no drag on the 1 mm diameter tow cable, which is made of Kevlar ($\rho = 1440$ kg/m^3, $\sigma_{ult} = 2760$ MPa). (a) Assuming that the towed body and the cable have no influence on the orbiter, what is the orbiter's velocity? (b) What is the towed body's velocity? (c) What is the drag force acting on the towed body? (d) What is the gravitational attraction force on the towed body? (e) What is the centrifugal force on the towed body? (f) What is the force exerted by the cable, on the towed body? (g) At what flight path angle does the cable act? (h) Assuming that the cable is straight, calculate its length. (i) What is the mass of the cable? (j) What is the stress in the cable considering only the force exerted by the towed body? (k) How much thrust must be produced by the orbiter, to maintain both, orbiter and towed body, on their respective circular orbits?

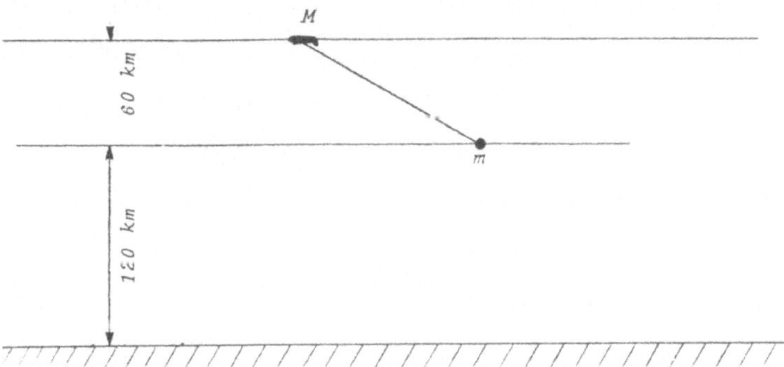

9.6   The shuttle orbiter happens to be loaded such that its mass amounts to $m = 100\,000$ kg. It circles the Earth at an altitude of 183 km, in locked rotation, at an attitude such that its frontal area assumes its largest value (460 m^2). If the orbiter loses 2.5 km altitude per orbit, determine the drag coefficient c_d.

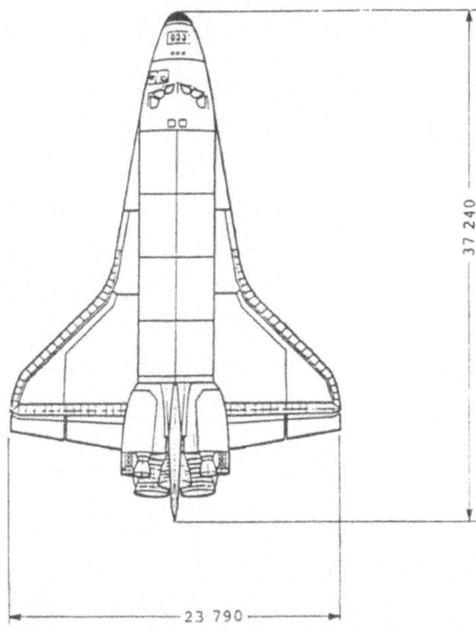

9.7   Engineers making plans about equipping the space shuttle with dipole antennas for plasma measurements, investigate a pair of antenna booms of 50 mm diameter and 150 m length each. If the
frontal area of the shuttle is 60 m^2 and the shuttle's drag coefficient is 3.9 (with cargo bay doors
open), and if the drag coefficient of the booms is 2, determine (a) the total drag force of the
shuttle-boom system at an altitude of 200 km above Earth surface. (b) How large is the force
transmitted from each boom to the shuttle, if the shuttle has a mass of 100 000 kg and each boom
a mass of 50 kg? (c) Calculate the tip deflection ($\delta = Pl^3/8EI$) due to atmospheric drag for an
elastic antenna boom with $EI = 1676\,\mathrm{Nm}^2$.

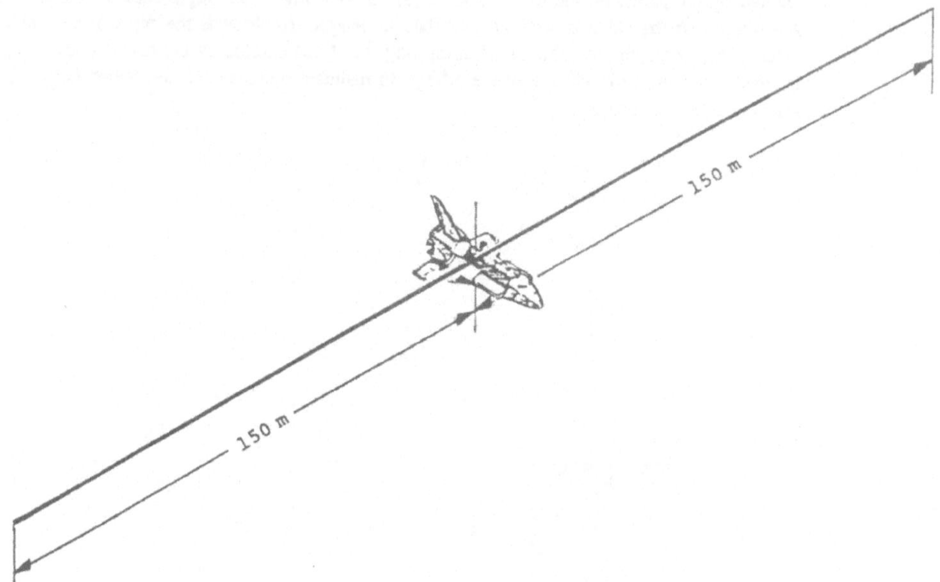

10

Satellites from Infinity

Consider a spacecraft that has left the Earth and is flying to Jupiter. Consider it launched such that it is going to encounter Mars on its way. Precautions are taken, of course, to make sure that it does not collide with Mars. On the other hand, it is to approach Mars closely, so close as a matter of fact, that it will be under the *influence* of Mars for a certain period of time.

During the flyby past Mars, the spacecraft becomes a satellite of Mars, so much so that other influences are essentially excluded. As far as the encounter with Mars is concerned the spacecraft appears from somewhere very far (almost infinity), flies past Mars, and departs again into the very far distance (almost infinity).

Well, let us go one step further and *approximate* the flyby manoeuvre. Let Mars *acquire* a satellite, which has come from infinity (as far as Mars is concerned), moves past Mars on a hyperbolic orbit, and leaves Mars again into infinity. We say that the spacecraft enters, and leaves again, the *influence sphere* of Mars. During the dwell in Mars' influence sphere, the spacecraft is a true (although not very faithful) satellite of Mars. We refer to the Sun as the spacecraft's *primary master*, and to Mars as the spacecraft's *secondary master*. The influence of the primary master upon the spacecraft is negligible, while the spacecraft is in the influence sphere of the secondary master.

The assumptions thus made lead to a patchwork of conics (i.e. the hyperbolic orbit about the secondary master is patched on to the lliptic orbits about the primary master). This *patched conics* approach provides a very good and useful first approximation for the spacecraft orbits. For more accurate orbit determinations, the influence of both masters must, of course, be taken into account simultaneously. Such orbit calculations must typically be carried out numerically.

Let us now look at the simple Kepler problem of a satellite on a hyperbolic orbit.

10.1 Hyperbolic Orbits

We consider a point satellite m (Figure 10.1) which is approaching its point (or spherical) master M from infinity at an *entrance velocity* (or *approach velocity*)

$$\mathbf{v}_{en}$$

whose magnitude is larger than zero.

We also assume that the influence sphere of the master either reaches as far as infinity, or is so large that the assumption is justified that the satellite is at infinity as it enters the influence sphere of its new master. Under these circumstances, we conclude that the magnitude of the entrance velocity is equal to the hyperbolic excess speed v_∞. The satellite will depart from the sphere of influence of the master at an infinite distance and that at an *exit velocity*

$$\mathbf{v}_{ex}$$

whose magnitude is also equal to the hyperbolic excess speed. We thus have

$$|\mathbf{v}_{ex}| \ = \ |\mathbf{v}_{en}| \ = \ v_\infty \tag{10.1}$$

In the case of a hyperbolic orbit, the speed equation (2.12) is

$$v \ = \ \sqrt{\mu\left[\frac{2}{r} + \frac{1}{a'}\right]} \tag{10.2}$$

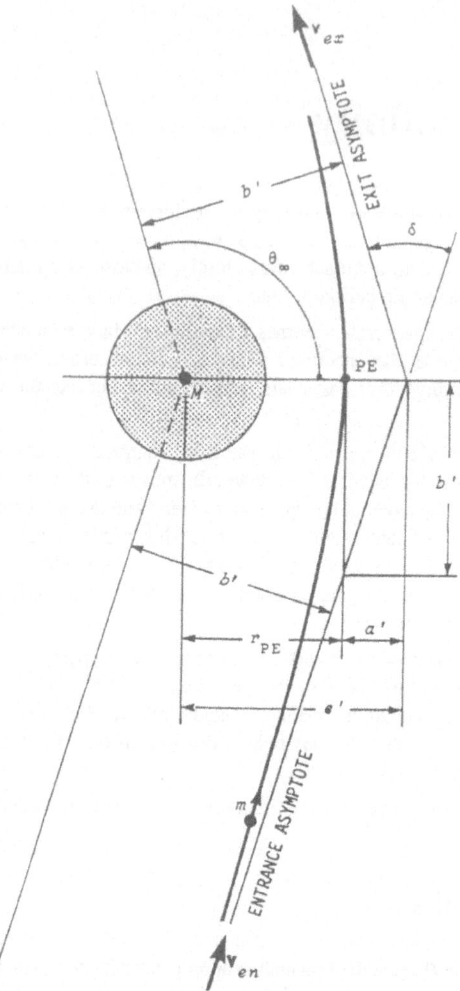

FIGURE 10.1. A satellite from infinity.

For $r = \infty$, the hyperbolic excess speed is obtained,

$$v_\infty = \sqrt{\frac{\mu}{a'}} \tag{10.3}$$

If $v_\infty = |v_{en}|$ is a known quantity, we can compute the semi-major axis of the orbit hyperbola

$$a' = \frac{\mu}{v_\infty^2} \tag{10.4}$$

Not only the entrance velocity's magnitude, but also its direction is known. From it, one can determine the distance b' at which the satellite would pass the master in case the latter would *not* exert any attractive force. The entrance velocity establishes the tangent of the orbit hyperbola at infinity, which we shall call the *entrance asymptote* (Figure 10.1).

With the aid of the equation

$$\tan\theta_\infty = -\frac{b'}{a'} = -\sqrt{\varepsilon^2 - 1} \tag{10.5}$$

the true anomaly of the entrance asymptote can be established. The *deflection angle* δ can be deduced from Figure 10.1. It is

$$\delta = 2\theta_\infty - 180° \tag{10.6}$$

Using equations (10.4) and (10.5), the deflection angle can also be expressed as

$$\delta = \pi - 2 \arctan\frac{b'v_\infty^2}{\mu} \tag{10.7}$$

The greater the magnitude of the entrance velocity and the greater the distance b' of the entrance asymptote, the smaller is the deflection angle.

The linear eccentricity, with the aid of which the location of the master is established (Figure 10.1), is

$$e' = a'\sqrt{1+\frac{b'^2}{a'^2}}$$

The numerical eccentricity is

$$\varepsilon = \sqrt{1+\frac{b'^2}{a'^2}} \tag{10.8}$$

Since, from equations (10.5) and (10.8)

$$\frac{1}{\varepsilon} = \frac{1}{\sqrt{1+\tan^2\theta_\infty}} = \cos\theta_\infty$$

and from equation (10.6)

$$\sin\frac{\delta}{2} = \sin(\theta_\infty - 90°) = \cos\theta_\infty$$

we find

$$\sin\frac{\delta}{2} = \frac{1}{\varepsilon} \tag{10.9}$$

Since $\varepsilon > 1$, we conclude that always $\delta < 180°$. In practice, however, the magnitude of the deflection angle δ will be considerably smaller than $180°$, because the finite size R of the master requires that $\varepsilon > 1 + \dfrac{R}{a'}$ (see equation (10.10)).

A satellite may not approach so close, that it collides with the finite size master or burns up in the master's atmosphere, in case there is an atmosphere. The approach between satellite and master is closest in the periapsis of the satellite's orbit. The distance between centre of master mass and periapsis must thus be larger than the master's radius R (including atmosphere if present)

$$r_{PE} = a'(\varepsilon - 1) > R \tag{10.10}$$

In the periapsis, the satellite also reaches its greatest velocity, viz.

$$v_{PE} = v_\infty\sqrt{\frac{\varepsilon+1}{\varepsilon-1}} \tag{10.11}$$

As an example, consider a spacecraft, approaching Mars from infinity. The entrance velocity has a magnitude of 4 km/s and a direction passing the centre of Mars at a distance of 9000 km. Any influence of the Mars trabants (Phobos and Deimos) is negligible. Find the semi-major axis, the true anomaly of the entrance asymptote, the deflection angle, the linear and the numerical eccentricity, the perimartian velocity and the perimartian distance of the orbit of the spacecraft (Mars: $\mu = 42\,828.3$ km^3/s^2). The semi-major axis is (Figure 10.1)

$$a' = \frac{\mu}{v_\infty^2} = \frac{42\,828.3}{4^2} = 2676.8 \text{ km}$$

For the true anomaly of the entrance asymptote,

$$\tan\theta_\infty = -\frac{b'}{a'} = -\frac{9000}{2676.8} = -3.362$$

giving

$$\theta_\infty \;=\; 106.565°$$

The orbit deflection angle is

$$\delta \;=\; 2\theta_\infty \;-\; 180° \;=\; 213.129° \;-\; 190° \;=\; 33.129°$$

The linear eccentricity is

$$e' \;=\; \sqrt{a'^2 + b'^2} \;=\; \sqrt{2676.8^2 + 9000^2} \;=\; 9389.6 \,\text{km}$$

The numerical eccentricity is

$$\varepsilon \;=\; \sqrt{1 + \frac{b'^2}{a'^2}} \;=\; \sqrt{1 + \frac{9000^2}{2676.8^2}} \;=\; 3.508$$

The perimartian velocity of the spacecraft is

$$v_{\text{PE}} \;=\; v_\infty \sqrt{\frac{\varepsilon + 1}{\varepsilon - 1}} \;=\; 4\sqrt{\frac{4.508}{2.508}} \;=\; 5.363 \,\text{km/s}$$

The distance between perimartian and Mars centre is

$$r_{\text{PE}} \;=\; a'(\varepsilon - 1) \;=\; 2676.8 \cdot 2.508 \;=\; 6712.8 \,\text{km}$$

The radius of Mars is 3415 km. The spacecraft will thus come as close to the Martian surface as

$$6712.8 \;-\; 3415 \;=\; 3297.8 \,\text{km}$$

10.2 The Collision Radius

The critical collision radius is that distance between centre of master and entrance asymptote of satellite, for which the satellite orbit osculates the surface of the master mass.

For osculation (Figure 10.2)

$$r_{\text{PE}} \;=\; R$$

where R is the outside radius of the master mass (assumed to be spherical). From equations (10.9), (10.10) and (10.7), we obtain for the collision radius b'_c

$$b'_c \;=\; \frac{v_{\text{PE}} R}{v_\infty} \qquad (10.12)$$

If one employs the escape velocity v_p from the surface of the master mass, the equation for the collision radius assumes the form

$$b'_c \;=\; R\sqrt{1 + \frac{v_p^2}{v_\infty^2}} \qquad (10.13)$$

or

$$b'_c \;=\; R\sqrt{1 + \frac{2\mu}{Rv_\infty^2}} \qquad (10.14)$$

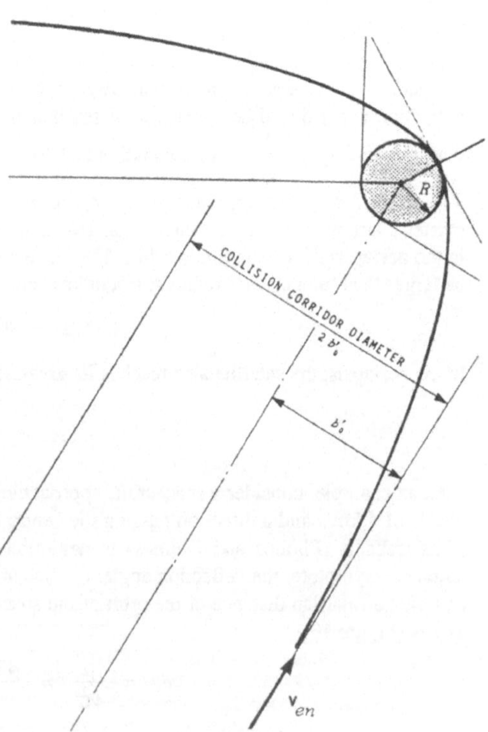

FIGURE 10.2. The collision radius and collision corridor.

If the satellite approaches its master on an entrance asymptote located closer to the centre of the master than the collision radius, then the point satellite will collide with the master mass.

If the satellite has an infinite approach velocity, $v_\infty = \infty$, then the collision radius becomes equal to the master radius. If the satellite approaches on a parabolic orbit, $v_\infty = 0$, then the collision radius becomes infinity.

The collision radius defines a *collision corridor* within which a satellite may not approach its master if a collision is to be avoided (Figure 10.2).

10.3 Primary Master and Secondary Master

In practical cases, the problem situation is frequently such that a satellite m (e.g. a spacecraft) of a primary master M_A (e.g. the Sun) enters temporarily into the influence sphere of a secondary master M_B (e.g. Mars). The secondary master is also a satellite of the primary master. The orbit of the satellite m about its primary master is changed because of the satellite's encounter with the secondary master. This change is to be determined.

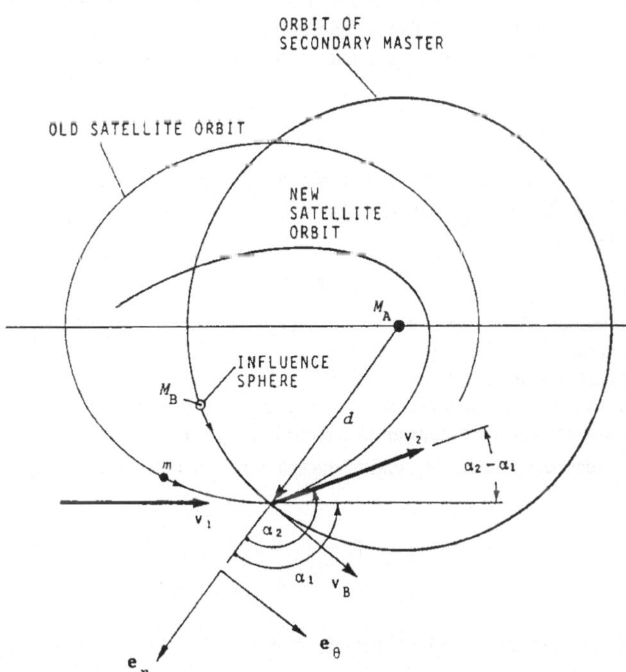

FIGURE 10.3. Primary master M_A, secondary master M_B, and satellite m.

In Figure 10.3, a satellite orbit about a primary master M_A is shown. The satellite is m. At point 1, the satellite enters the influence sphere of a secondary master M_B. The secondary master M_B is a satellite of the primary master on an orbit also shown in Figure 10.3.

A very good first approximation of the satellite behaviour is obtained by making the following assumptions:

1. The influence sphere of the secondary master, in comparison with the size of the orbit of the satellite, is so small, i.e.

$$r_{IS} \ll |\mathbf{d}| \tag{10.15}$$

that the influence sphere can be considered to be of point size on the orbit of the satellite: The entrance point 1, the position of the secondary master, and the exit point 2 can be assumed to coincide (see e.g. Figure 10.7).

2. Compared to the size R of the secondary master, however, the influence sphere of the secondary master is so large, that the relative orbit of the satellite with respect to the secondary master can be represented by a hyperbola (Figure 10.4), on which the satellite, coming from infinity, approaches the secondary master.

$$R \;\ll\; r_{\text{IS}} \tag{10.16}$$

3. The dwell period of the satellite in the influence sphere of the secondary master is so short, in comparison with the satellite's orbital period about its primary master, that the velocity v_B and the position $\mathbf{d}$ of the secondary master change so little, that one can set

$$\mathbf{d} \;=\; \text{constant}$$

and

$$v_B \;=\; \text{constant} \tag{10.17}$$

during flyby.

At the moment of entry into the influence sphere of the secondary master, the satellite has an absolute velocity, which we shall designate $\mathbf{v}_1$. The secondary master itself has an absolute velocity $\mathbf{v}_B$. The entrance velocity $\mathbf{v}_{en}$ of the satellite into the influence sphere of the secondary master is a relative velocity and is given by

$$\mathbf{v}_{en} \;=\; \mathbf{v}_1 \;-\; \mathbf{v}_B \tag{10.18}$$

The exit velocity $\mathbf{v}_{ex}$ has the same magnitude as the entrance velocity, but is deflected by an angle δ, which can be determined by means of equation (10.6). The satellite leaves the influence sphere of the secondary master at an absolute velocity

$$\mathbf{v}_2 \;=\; \mathbf{v}_B \;+\; \mathbf{v}_{ex} \tag{10.19}$$

An accurate determination of the new velocity is a very protracted process. Therefore, we make use of the approximate method outlined above, which is very well suited for the prediction of spacecraft behaviour in a Sun-planet system, e.g. in the Sun-Mars system. Because of the relatively large mass of the Moon in comparison to the Earth, the approximate method outlined above can unfortunately only give a very rough estimate of spacecraft behaviour in the Earth-Moon system.

The positions of the secondary master M_B and of the spacecraft m with respect to the primary master M_A are both approximately

$$\mathbf{r} \;=\; \begin{bmatrix} \mathbf{e}_r \, \mathbf{e}_\theta \, \mathbf{e}_z \end{bmatrix} \begin{bmatrix} d \\ 0 \\ 0 \end{bmatrix}$$

where $\mathbf{e}_r$ and $\mathbf{e}_\theta$ are radial and transverse basis vectors for orbits about the primary master M_A.

Thus, at entry, their radial velocity components are almost collinear, and same can be said about their transverse velocity components. Because of equations (10.1) and (10.18), we can then obtain the magnitude of the entry velocity (Figure 10.4) as

$$v_\infty \;=\; \sqrt{(v_{1_r} - v_{B_r})^2 \;+\; (v_{1_\theta} - v_{B_\theta})^2}$$

If the secondary master moves on a circular orbit, then $v_{B_r} = 0$ and $v_{B_\theta} = v_B$, and consequently

$$v_\infty \;=\; \sqrt{v_{1_r}^2 \;+\; (v_B - v_{1_\theta})^2} \tag{10.20}$$

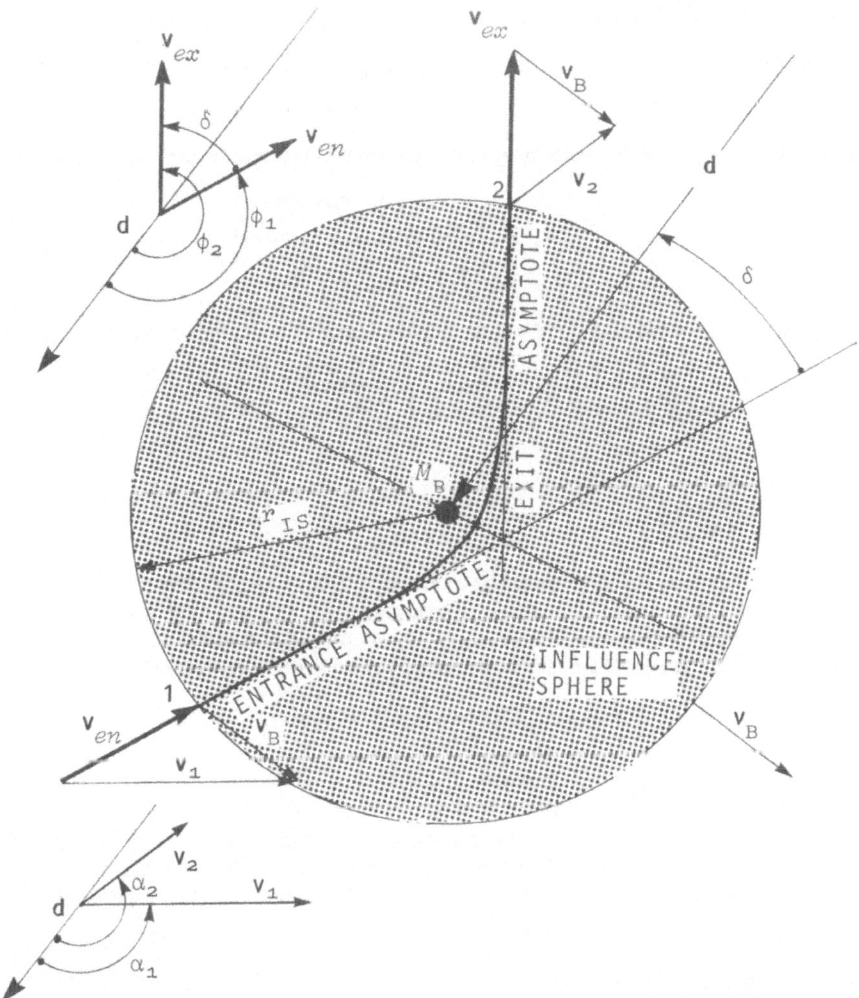

FIGURE 10.4. Entry into and exit out of the influence sphere of the secondary master.

The direction of the $\mathbf{v}_{en}$ vector, which respect to $\mathbf{d}$, is obtained from

$$(\mathbf{v}_1 - \mathbf{v}_B) \cdot \mathbf{d} = v_\infty d \cos \phi_1$$

or

$$\cos\phi_1 = \frac{(\mathbf{v}_1 - \mathbf{v}_B) \cdot \mathbf{d}}{v_\infty d} = \frac{v_{1_r}}{v_\infty} \tag{10.21}$$

The exit velocity is, with $\phi_2 = \phi_1 + \delta$ (Figure 10.4),

$$\mathbf{v}_{ex} = \begin{bmatrix} \mathbf{e}_r \, \mathbf{e}_\theta \, \mathbf{e}_z \end{bmatrix} \begin{bmatrix} v_\infty \cos \phi_2 \\ v_\infty \sin \phi_2 \\ 0 \end{bmatrix}$$

Care must be taken to establish whether the deflection angle δ is positive (Figure 10.4) or negative in relation to the angle ϕ.

The secondary master's velocity is

$$v_B = \begin{bmatrix} e_r \, e_\theta \, e_z \end{bmatrix} \begin{bmatrix} 0 \\ v_B \\ 0 \end{bmatrix}$$

The sum $v_B + v_{ex}$ is the new velocity v_2 of the spacecraft with respect to the primary master

$$v_2 = \begin{bmatrix} e_r \, e_\theta \, e_z \end{bmatrix} \begin{bmatrix} v_\infty \cos \phi_2 \\ v_B + v_\infty \sin \phi_2 \\ 0 \end{bmatrix} \tag{10.22}$$

The angle α_2 between radius vector d and spacecraft velocity v_2 can be obtained from

$$\cos \alpha_2 = \frac{v_\infty}{v_2} \cos \phi_2 \tag{10.23}$$

10.4 Energy Gain

A satellite (of a primary master) experiences a change in velocity, magnitude as well as direction, during an encounter with a secondary master. This flyby manoeuvre can, e.g., lead to an increase of the magnitude of the velocity, which in turn leads to an energy gain and that without having to carry an on-board energy source. The kinetic energy of the satellite during a flyby is increased by an amount

$$\Delta T = \frac{1}{2} m v_2^2 - \frac{1}{2} m v_1^2 \tag{10.24}$$

while the potential energy remains constant (because $d = $ constant). The energy may, of course, also experience a decrease during the flyby, viz. when $v_2 < v_1$.

An energy gain is taken advantage of in case of the so-called *Grand Tour*, where a spacecraft from Earth to e.g. Uranus is launched in one of the (rare!) years where Mars, Jupiter, and Saturn are located such that each influence sphere is traversed by the spacecraft, in such manner that energy is picked up each time.

10.5 Influence Sphere

The exact size of the influence sphere used in the preceding derivations is of little concern when the assumptions used in the present chapter are made. Furthermore, there are several ways of defining an influence sphere. On the other hand, one might want to have some idea as to the relative size of an influence sphere and we shall therefore establish a definition for an influence sphere in the following. The definition is based on the condition that the ratios of perturbation acceleration to unperturbed acceleration for a spacecraft with respect to its primary master and it its secondary master are equal.

We consider a primary master M_A and a secondary master M_B, which itself is a satellite of the master M_A. The mass m of the spacecraft is much smaller than that of the secondary master, which in turn is much smaller than that of the primary master.

$$m \ll M_B \ll M_A \tag{10.25}$$

The spacecraft m is located at a position r measured from the secondary master, and at a position R measured from the primary master (Figure 10.5).

10.6 The Spacecraft as Satellite of the Secondary Master

In Figure 10.5, three masses are shown. The spacecraft m is positioned close to its secondary master M_B.

For the absolute acceleration $\mathbf{a}_m$ of the spacecraft, we can write

$$\mathbf{a}_m = \mathbf{a}_B + \mathbf{a}_{m/B} \tag{10.26}$$

The relative acceleration $\mathbf{a}_{m/B}$ of the spacecraft with respect to the secondary master is equal to $\ddot{\mathbf{r}}$.

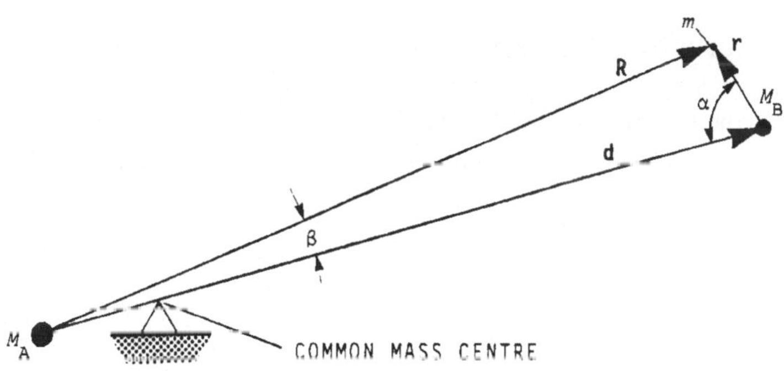

FIGURE 10.5. A spacecraft m with primary and secondary master.

The attraction force exerted by the primary master on the secondary master causes an absolute acceleration of the secondary master of

$$\mathbf{a}_B = -\frac{\mu_A}{d^3}\mathbf{d} \tag{10.27}$$

On the other hand, the absolute acceleration $\mathbf{a}_m$ of the spacecraft is, according to the law of gravitational attraction,

$$\mathbf{a}_m = -\frac{\mu_B}{r^3}\mathbf{r} - \frac{\mu_A}{R^3}\mathbf{R} \tag{10.28}$$

Summing up and solving for the relative acceleration of the spacecraft with respect to the secondary master

$$\ddot{\mathbf{r}} = -\frac{\mu_B}{r^3}\mathbf{r} - \frac{\mu_A}{R^3}\mathbf{R} + \frac{\mu_A}{d^3}\mathbf{d} \tag{10.29}$$

or

$$\ddot{\mathbf{r}} = -\frac{\mu_B}{r^3}\mathbf{r} - \frac{\mu_A}{R^3}\left[\mathbf{R} - \frac{R^3}{d^3}\mathbf{d}\right] \tag{10.30}$$

If one compares equation (10.22) with the situation where a satellite m is under the influence of a master M_B without any perturbing influence from any other celestial body, then the term

$$-\frac{\mu_B}{r^3}\mathbf{r}$$

represents the unperturbed portion of the acceleration. The presence of the primary master M_A is the cause of a *perturbation acceleration*

$$\ddot{\mathbf{r}}_{pert} \;=\; -\frac{\mu_A}{R^3}\left[\mathbf{R} - \frac{R^3}{d^3}\mathbf{d}\right] \tag{10.31}$$

such that

$$\ddot{\mathbf{r}} \;=\; -\frac{\mu_B}{r^3}\mathbf{r} \;+\; \ddot{\mathbf{r}}_{pert} \tag{10.32}$$

10.7 Perturbation Acceleration

The magnitude of the perturbation acceleration can be determined with the aid of a dot product.
We introduce temporarily the angles α and β (Figure 10.5).

$$\ddot{\mathbf{r}}_{pert} \cdot \ddot{\mathbf{r}}_{pert} \;=\; \frac{\mu_A^2}{d^4}\left[1 + \frac{d^4}{R^4} - \frac{2d^2}{R^2}\cos\beta\right] \tag{10.33}$$

Since

$$R\,\cos\beta \;+\; r\,\cos\alpha \;=\; d \tag{10.34}$$

one can also write

$$\ddot{\mathbf{r}}_{pert} \cdot \ddot{\mathbf{r}}_{pert} \;=\; \frac{\mu_A^2}{d^4}\left[1 + \frac{d^4}{R^4} - 2\frac{d^3}{R^3} + \left(2\frac{r}{d}\cos\alpha\right)\frac{d^3}{R^3}\right] \tag{10.35}$$

The cosine law applied to the triangle in Figure 10.5 gives

$$R^2 \;=\; d^2 \;+\; r^2 \;-\; 2rd\,\cos\alpha \tag{10.36}$$

or

$$\frac{R}{d} \;=\; \sqrt{1 - \left[2\frac{r}{d}\cos\alpha - \frac{r^2}{d^2}\right]} \tag{10.37}$$

We now limit the analysis to a spacecraft positioned near the secondary master M_B, i.e. we require $r \ll d$. We expand equation (10.37) into binomial series for $(R/d)^{-4}$ and $(R/d)^{-3}$, each of which we break off after the term for r^2/d^2. The result is

$$\left[\frac{R}{d}\right]^{-4} \;=\; 1 \;+\; 4\frac{r}{d}\cos\alpha \;-\; 2\frac{r^2}{d^2} \;+\; 12\frac{r^2}{d^2}\cos^2\alpha \tag{10.38}$$

and

$$\left[\frac{R}{d}\right]^{-3} \;=\; 1 \;+\; 3\frac{r}{d}\cos\alpha \;-\; \frac{3}{2}\frac{r^2}{d^2} \;+\; \frac{15}{2}\frac{r^2}{d^2}\cos^2\alpha \tag{10.39}$$

These expansions entered into equation (10.35) result in

$$|\ddot{\mathbf{r}}_{pert}| \;=\; \frac{\mu_A}{d^3}\sqrt{1 + 3\cos^2\alpha} \tag{10.40}$$

10.8 The Spacecraft as a Satellite of the Primary Master

We can derive the relative acceleration $\ddot{\mathbf{R}}$ of the spacecraft with respect to the primary master in a similar fashion, and obtain

$$\ddot{\mathbf{R}} = -\frac{\mu_A}{R^3}\mathbf{R} - \frac{\mu_B}{r^3}\left[\mathbf{r} + \frac{r^3}{d^3}\mathbf{d}\right] \tag{10.41}$$

The perturbation acceleration

$$\ddot{\mathbf{R}}_{pert} = -\frac{\mu_B}{r^3}\left[\mathbf{r} + \frac{r^3}{d^3}\mathbf{d}\right] \tag{10.42}$$

has a magnitude of

$$|\ddot{\mathbf{R}}_{pert}| = \sqrt{\ddot{\mathbf{R}}_{pert} \cdot \ddot{\mathbf{R}}_{pert}} = \frac{\mu_B}{r^2}\sqrt{1 + \frac{r^4}{d^4} - (2\cos\alpha)\frac{r^2}{d^2}} \tag{10.43}$$

Setting again $r \ll d$, one gets

$$|\ddot{\mathbf{R}}_{pert}| = \frac{\mu_B}{r^2} \tag{10.44}$$

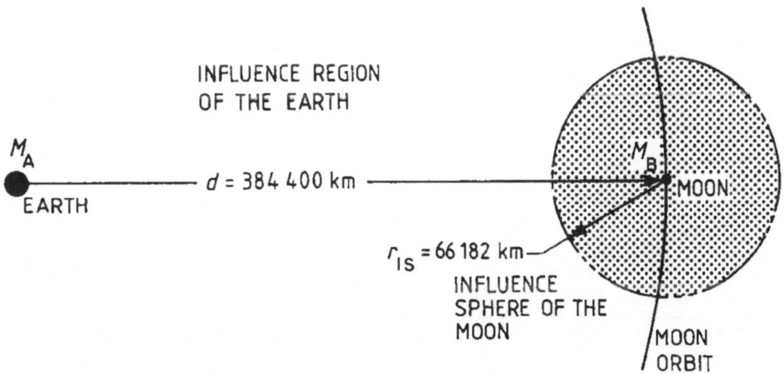

FIGURE 10.6. Influence sphere of the Moon in the Earth-Moon system.

10.9 The Tisserand Equation

We had already noted that there are several definitions for the size of the influence sphere of a secondary master. We shall consider the condition proposed by Tisserand[*] (1889) and follow his derivation of the radius of the influence sphere. Tisserand's measure for the limit of the influence of a master in relation to the influence of another master is that the ratios of the perturbed to unperturbed accelerations of a spacecraft with respect to one master and with respect to the other master are equal. In form of an equation

$$\frac{|\ddot{\mathbf{r}}_{pert}|}{\dfrac{\mu_B}{r^2}} = \frac{|\ddot{\mathbf{R}}_{pert}|}{\dfrac{\mu_A}{R^2}} \tag{10.45}$$

---

[*] François Felix Tisserand (1845-1896), French astronomer.

Making use of equations (10.40) and (10.44), and solving for the distance from the secondary master

$$r^5 = \frac{\mu_B^2}{\mu_A^2} \frac{d^3 R^2}{\sqrt{1 + 3\cos^2\alpha}} \tag{10.46}$$

If $r << d$, then $R \approx d$ as a consequence (Figure 10.8), and

$$r = \left[\frac{\mu_B}{\mu_A}\right]^{2/5} \frac{d}{(1 + 3\cos^2\alpha)^{0.1}} \tag{10.47}$$

Finally one can go one step further in a simplification process and replace $(1 + 3\cos^2\alpha)^{0.1}$ by unity. After all, the root can only vary between unity (when $\cos\alpha = 0$) and $4^{0.1} = 1.149$ (when $\cos\alpha = 1$). Proceeding in this fashion, one obtains the *Tisserand equation* for the limit radius of a now spherical influence zone

$$r_{IS} = \frac{d}{\left[\dfrac{\mu_A}{\mu_B}\right]^{0.4}} \tag{10.48}$$

For the influence sphere of the Moon (Figure 10.6) in the Earth-Moon system (Earth: $\mu_A = 398\,601.19\ \text{km}^3/\text{s}^2$, Moon: $\mu_B = 4902.73\ \text{km}^3/\text{s}^2$, $d = 384\,400\ \text{km}$)

$$r_{IS} = 66\,182\ \text{km} \tag{10.49}$$

For the influence sphere of Mars in the Sun-Mars system (Sun: $\mu_A = 132\,712\,520\,000\ \text{km}^3/\text{s}^2$, Mars: $\mu_B = 42\,828.3\ \text{km}^3/\text{s}^2$, mean distance between Sun and Mars: $d = 228 \cdot 10^6\ \text{km}$)

$$r_{IS} = 0.577 \cdot 10^6\ \text{km} \tag{10.50}$$

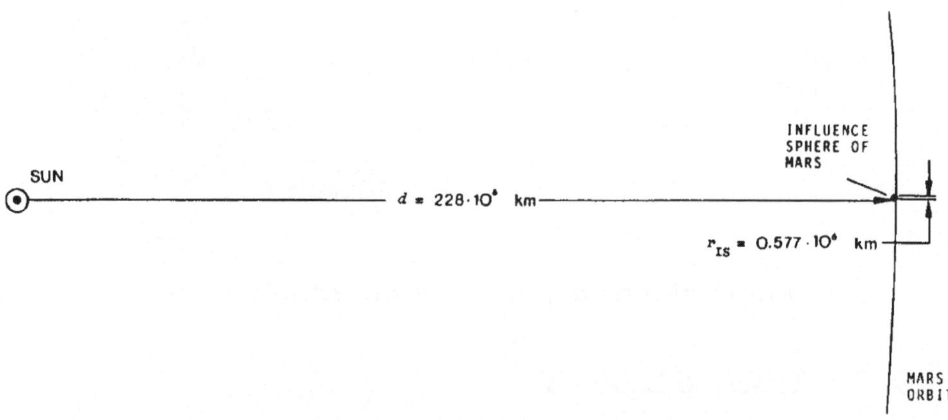

FIGURE 10.7. The influence sphere of Mars in the Sun-Mars system.

In case of the Sun-Mars system, the influence sphere of Mars (Figure 10.7) appears indeed to be of point size.

$$\frac{r_{IS}}{d} = \frac{0.577}{288} = 0.002\,530$$

Mars and the influence sphere of Mars are shown in Figure 10.8.

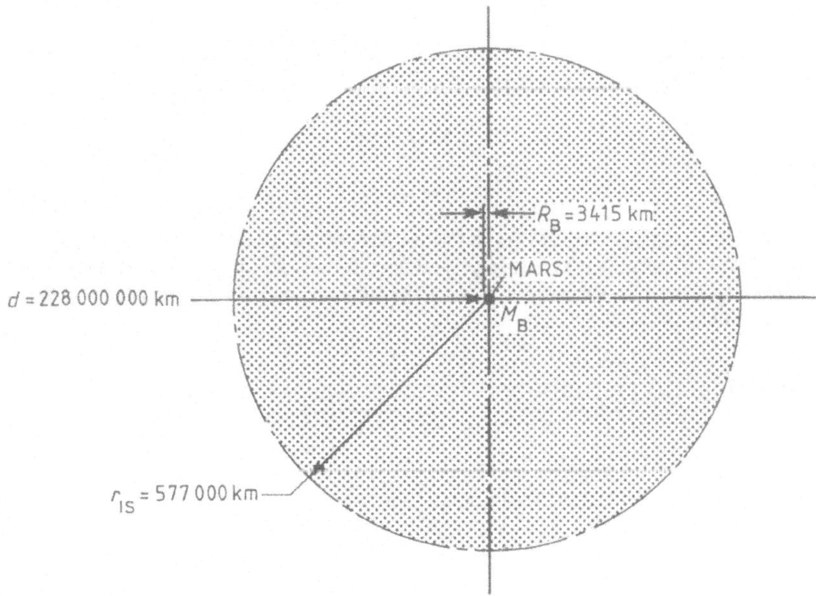

FIGURE 10.8. Mars and its influence sphere.

For the influence sphere of the Earth in the Sun-Earth system (Sun: μ_A = 132 712 520 000 km^3/s^2, Earth: μ_B = 398 601.19 km^3/s^2, mean distance between Sun and Earth: d = 150 $\cdot$ 10^6 km)

$$r_{IS} = 0.927 \cdot 10^6 \text{ km} \tag{10.51}$$

The Moon's orbit with a mean radius of 384 400 km lies thus well within the Earth's influence sphere, with its radius of 927 000 km.

We can now also obtain some insight into the relative magnitude of the perturbation portion of the acceleration of a spacecraft. With the aid of equations (10.40) and (10.47), we find at the boundary of the influence sphere

$$\frac{|\ddot{\mathbf{r}}_{pert}|_{IS}}{\frac{\mu_B}{r_{IS}^2}} = \frac{(1 + 3\cos^2\alpha)^{0.2}}{\left[\frac{\mu_A}{\mu_B}\right]^{0.2}} \tag{10.52}$$

We replace $(1 + 3\cos^2\alpha)^{0.2}$ by unity, solve for the ratio of perturbation acceleration to unperturbed acceleration, and obtain

$$\frac{|\ddot{\mathbf{r}}_{pert}|_{IS}}{\frac{\mu_B}{r_{IS}^2}} = \frac{1}{\left[\frac{\mu_A}{\mu_B}\right]^{0.2}} \tag{10.53}$$

For a spacecraft at the boundary of the Earth's influence sphere in the Sun-Earth system, this ratio is e.g. 0.079, i.e. the perturbation portion represents 7.9% of the theoretical unperturbed portion of the satellite's acceleration right at the boundary of the influence sphere. Within the influence sphere, the perturbation portion is, of course, even smaller. Generally speaking, the perturbation portion within an influence sphere is always

$$|\ddot{\mathbf{r}}_{pert}| < \left[\frac{\mu_B}{\mu_A}\right]^{0.2} \frac{\mu_B}{r_{IS}^2} \tag{10.54}$$

Suggested Reading

1.   Tisserand, F. *Traité de mécanique céleste*, Gauthier-Villars, Paris, 1889, Vol. I, 474 pp.; Vol. II, 552 pp.; Vol. III, 427 pp.; Vol. IV, 548 pp.

Problems

10.1   What is the radius r_{IS} of the influence sphere of (a) Mercury ($\mu_B = 2.2 \cdot 10^4$ km^3/s^2, $d = 58 \cdot 10^6$ km); (b) Venus ($\mu_B = 3.25 \cdot 10^5$ km^3/s^2, $d = 108 \cdot 10^6$ km); (c) Earth; (d) Jupiter ($\mu_B = 1.3 \cdot 10^8$ km^3/s^2, $d = 778 \cdot 10^6$ km); and (e) Saturn ($\mu_B = 3.8 \cdot 10^7$ km^3/s^2, $d = 1427 \cdot 10^6$ km).

10.2   Given is

$$y = \sqrt{1 - (2ax - x^2)}$$

where $|x| \ll 1$ and $-1 < a < 1$.
By means of a binomial expansion, find (a) $y^{-3} = $ ____ ; and (b) $y^{-4} = $ ____ up to and including all terms in x^2.

10.3   A spacecraft approaches Mars ($\mu = 42\,828.3$ km^3/s^2, $R = 3383$ km) form infinity at a velocity which is directed such that it has a distance of 8000 km from Mars centre. If the spacecraft is to approach Mars just outside the collision corridor, what must (a) its hyperbolic excess speed be? (b) What is the angle between exit velocity and entrance velocity? (c) *Estimate roughly* for how long the spacecraft will be within the influence sphere of Mars ($r_{IS} = 577\,000$ km); and (d) *calculate* for how long the satellite will be within the influence sphere of Mars.

10.4   A spacecraft ($m = 800$ kg) on a Hohmann transfer orbit from the Earth ($a = 149.6 \cdot 10^6$ km) to Saturn ($a = 1427 \cdot 10^6$ km), flies unexpectedly through the influence sphere of Jupiter ($a = 778.3 \cdot 10^6$ km). The spacecraft approaches Jupiter ($\mu = 1.267 \cdot 10^8$ km^3/s^2) on an entry asymptote offset by $b' = 900\,000$ km. Assume circular coplanar orbits for Earth, Jupiter, and Saturn (Sun: $\mu = 1.327 \cdot 10^{11}$ km^3/s^2). (a) What will be the spacecraft's energy gain, if it passes ahead of Jupiter? (b) What will be the spacecraft's energy gain, if it passes behind Jupiter?

10.5   A spacecraft is launched from the Earth's orbit (150 Gm radius) and is to reach Jupiter's orbit (778 Gm radius). Perihel of the spacecraft's orbit is to be on the Earth's orbit, aphel on Jupiter's. Assume that all three, Earth, Mars, and Jupiter, orbit the Sun on coplanar circular paths. (a) Compute the elements a, ε, b, p, τ_T of the spacecraft's orbit and make a sketch. (b) Where (r, θ) does the spacecraft cross the orbit (228 Gm radius) of Mars? (c) What is the spacecraft's velocity when it encounters Mars' orbit? (d) What are the magnitudes of the radial and the transverse components of the spacecraft's velocity at the moment of encounter? If the spacecraft intersects Mars' orbit just at a time when the planet itself happens to be at the intersection point, then Mars will appreciably influence the path of the spacecraft. The spacecraft is to enter Mars' influence sphere and subsequently to pass the centre of Mars as close as 4000 km. Further, Mars and spacecraft are to orbit the Sun in the same sense, and the spacecraft is to overtake Mars on the side away from the Sun. (e) What is Mars' velocity? (f) What is the spacecraft's relative velocity, as it enters Mars' influence sphere? (g) Compute the elements a', b', θ_∞, of the spacecraft's orbit relative to Mars, within the influence sphere. (h) Compute the orbit deflection angle δ. (i) What is the spacecraft's relative velocity of exit from the influence sphere, and what is its absolute velocity at exit? (j) How much energy did the spacecraft ($m = 500$ kg) gain as a consequence of the encounter with Mars? (k) How long did it take the spacecraft to transverse Mars' influence sphere ($r_{IS} = 577\,000$ km)? (l) What are the elements a_2, p_2, ε_2, b_2, θ_2 of the spacecraft's new orbit about the Sun? Make a sketch. (m) Compute the absolute deflection angle $\alpha_2 - \alpha_1$. (n) By how much will the spacecraft exceed its aim (i.e. Jupiter's orbit), as a consequence of the encounter with Mars? (o) What distance did Mars cover during the spacecraft's dwell in Mars' influence spheres? (p) When (i.e. how long after launch from the Earth) will the spacecraft encounter Jupiter's orbit?

10.6 A spacecraft enters the Martian influence sphere with a hyperbolic excess speed of 1.2 km/s, directed such that it is 20 000 km from Mars centre. In the periapsis of its orbit about Mars, the spacecraft's speed is suddenly reduced by 0.5 km/s. Calculate (a) the semi-major axis, (b) the eccentricity of the subsequent orbit, and (c) make a sketch.

10.7 One theory has it, that some $3.5 \cdot 10^9$ years (3.5 billion years) ago the Moon, on an orbit about the Sun similar to that of the Earth, was captured by the Earth. This capture can be explained by huge tidal waves (change of gravitational potential of Earth, energy dissipation) on the Earth, and/or the impact of large bodies with the Moon (which formed the mare basins). Assume that the Moon entered the Earth's influence sphere at an hyperbolic excess speed of 0.1 km/s, such that the perigee of its orbit would be at 384 400 km from Earth centre. (a) What is the Δv that must have been imparted to the Moon by a body hitting the Moon in its perigee to ensure capture of the Moon, (b) to put the Moon into an elliptic orbit with $r_{PE} = 484\,400\text{km}$ and $r_{AP} = 700\,000$ km, and (c) to put the Moon into a circular orbit?

10.8 A spacecraft is to become a satellite of Venus ($R = 6114$ km, $\mu = 324\,860\text{ km}^3/\text{s}^2$). The spacecraft approaches from infinity, with a hyperbolic excess speed of 1.5 km/s, directed such that it is 45 000 km from Venus centre. (a) What velocity kick must be applied in the periapsis point to make the spacecraft move onto a circular orbit about Venus? (b) What is the radius of the new orbit?

10.9 Halley's comet has a period of 76.0 years. (a) Find the length of the comet orbit's semi-major axis. The orbit's eccentricity is 0.967. Find (b) its perihelion and (c) its aphelion, as well as (d) its semi-minor axis. Sketch (e) Halley's orbit, Neptune's (cicular) orbit ($r = 4\,497 \cdot 10^6$ km), and the Earth's (circular) orbit around the Sun. (f) Noting that Halley's comet moves, say, clockwise, while the Earth moves counterclockwise around the Sun, sketch and calculate at which true anomaly of Halley's orbit an encounter with the Earth might take place, assuming that both moved in one and the same orbital plane. (g) What would be the relative entry velocity into the Earth's influence sphere? (h) By how much would the relative velocity of Halley's comet be deflected while passing through the Earth's influence sphere, if the entry asymptote's offset was $b' = 20\,000$ km? (i) What is the influence of Halley's comet ($m = 2 \cdot 10^{14}$ kg) on the orbit of the Earth ($m = 5.976 \cdot 10^{24}$ kg) during the encounter?

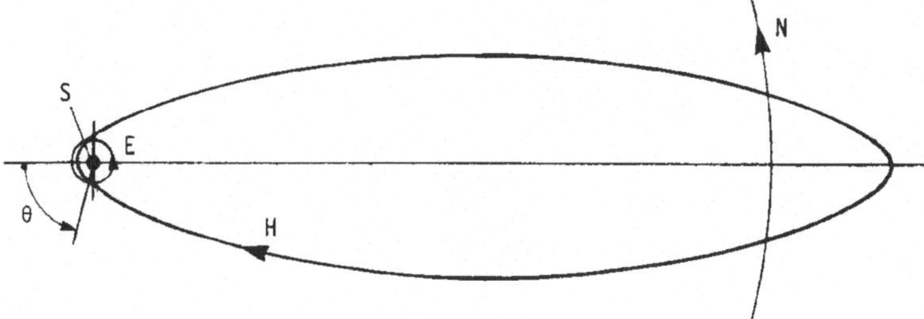

10.10 A spacecraft (m = 200 kg) is launched from Earth into an elliptic orbit about the Sun. At a true anomaly of 90°, it encounters Mars, such that its approach asymptote is offset by 4000 km, and that it passes behind Mars. (a) What is the spacecraft's kinetic energy gain due to the flyby? (b) Calculate the collision radius, and check whether the spacecraft will collide with Mars. Assume that the Earth and Mars are on circular orbits, and that the Earth, Mars, and spacecraft orbits are coplanar.

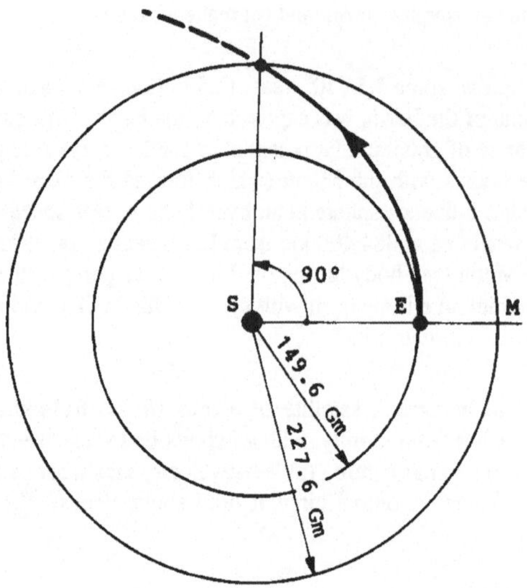

11

The General Two-Body Problem

In case the mass m of the satellite is so large that it can no longer be looked upon as being negligible in relation to the mass M of the master, the assumption that the satellite has only a negligible influence upon the master can no longer be maintained. One of the few systems of the universe for which this simplifying assumption e.g. cannot be made very often for practical cases is the Earth-Moon system, because of the relatively large mass of the Moon.

For an analysis of the general two-body problem, we inspect two point masses M and m, which attract each other by a force of magnitude

$$K = \frac{GMm}{r^2} \tag{11.1}$$

The point mass M is located at $\mathbf{r}_M$, the point mass m is located at $\mathbf{r}_m$. The common mass centre C is located at $\mathbf{r}_C$ (Figure 11.1). The origin 0 of the vectors is "fixed" in space, i.e. its acceleration is zero.

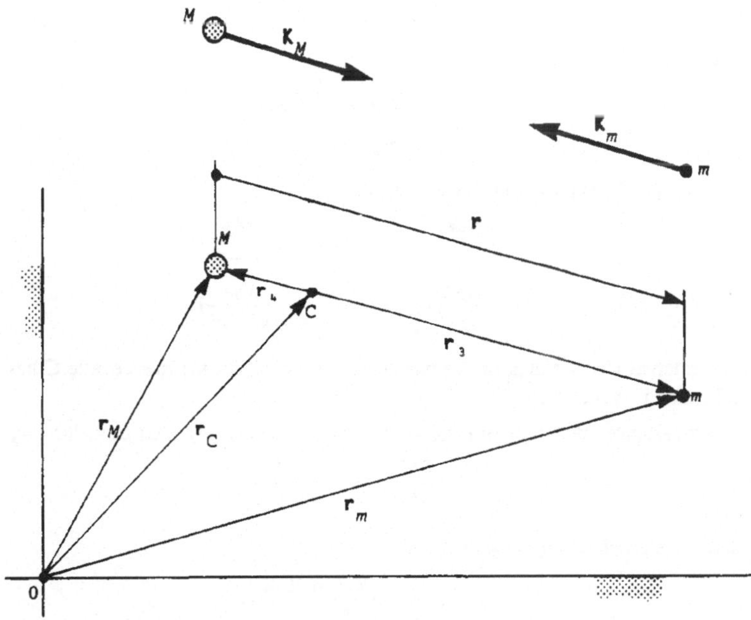

FIGURE 11.1. Two masses and their common mass centre C.

The position vector $\mathbf{r}$ originates at mass M and ends at mass m, such that

$$\mathbf{r} = \mathbf{r}_m - \mathbf{r}_M \tag{11.2}$$

Using the common mass centre C as origin, we find mass M located at

$$\mathbf{r}_4 = -\frac{m}{M+m}\mathbf{r} \tag{11.3}$$

and mass m located at

$$r_3 = \frac{M}{M+m}\mathbf{r} \tag{11.4}$$

The position vectors for the two masses can now be expressed as

$$\mathbf{r}_M = \mathbf{r}_C - \frac{m}{M+m}\mathbf{r} \tag{11.5}$$

$$\mathbf{r}_m = \mathbf{r}_C + \frac{M}{M+m}\mathbf{r} \tag{11.6}$$

The corresponding accelerations are

$$\ddot{\mathbf{r}}_M = \ddot{\mathbf{r}}_C - \frac{m}{M+m}\ddot{\mathbf{r}} \tag{11.7}$$

$$\ddot{\mathbf{r}}_m = \ddot{\mathbf{r}}_C + \frac{M}{M+m}\ddot{\mathbf{r}} \tag{11.8}$$

The force acting on mass M is

$$\mathbf{K}_M = \frac{GMm}{r^2}\mathbf{e}_r \tag{11.9}$$

The force acting on mass m is

$$\mathbf{K}_m = -\frac{GMm}{r^2}\mathbf{e}_r \tag{11.10}$$

According to Newton's second law

$$\mathbf{K}_M = M\ddot{\mathbf{r}}_M \tag{11.11}$$

$$\mathbf{K}_m = m\ddot{\mathbf{r}}_m \tag{11.12}$$

or with equations (11.9), (11.10), (11.7) and (11.8)

$$\frac{GMm}{r^2}\mathbf{e}_r = M\ddot{\mathbf{r}}_C - \frac{Mm}{M+m}\ddot{\mathbf{r}} \tag{11.13}$$

$$\frac{GMm}{r^2}\mathbf{e}_r = m\ddot{\mathbf{r}}_C + \frac{Mm}{M+m}\ddot{\mathbf{r}} \tag{11.14}$$

There are no external forces acting on the two-mass system. Thus, the mass centre C has no acceleration, i.e. its acceleration $\ddot{\mathbf{r}}_C$ vanishes.

Let us now investigate the motion of mass m. From equation (11.14), after cancelling m,

$$\frac{GM}{r^2}\mathbf{e}_r = -\frac{M}{M+m}\ddot{\mathbf{r}} \tag{11.15}$$

and for the relative acceleration, we obtain

$$\ddot{\mathbf{r}} = -\frac{M+m}{M}\frac{GM}{r^2}\mathbf{e}_r \tag{11.16}$$

We introduce a *gravitational parameter* μ_1 for the two-mass system

$$\mu_1 = G(M+m) = GM\left(1+\frac{m}{M}\right) = \mu\left(1+\frac{m}{M}\right) \tag{11.17}$$

and obtain for the relative acceleration of mass m with respect to mass M

$$\ddot{\mathbf{r}} = -\frac{\mu_1}{r^2}\mathbf{e}_r \tag{11.18}$$

an expression identical to equation (1.19) with one exception: The gravitational parameter is now $\mu' = \mu_M + \mu_m$, instead of $\mu = \mu_M$.

The world's first manned space station "Saljut", with spacecraft "Sojus 11" docked (June 1971), represents a relatively heavy artificial satellite. It weighed around 25 tonnes ($\approx 25 \cdot 10^3$ kg). The Earth weighs $5.976 \cdot 10^{24}$ kg. The gravitational parameter for this Earth-satellite system is thus

$$\mu' = \mu \left[1 + \frac{m}{M} \right] = \mu \left[1 + \frac{25 \cdot 10^3}{5.976 \cdot 10^{24}} \right] \tag{11.19}$$

or

$$\mu' = (1 + 4.183 \cdot 10^{-21}) \mu \tag{11.20}$$

Truly justification enough for neglecting m/M for artificial satellites and using the simplified two-body problem theory!

11.1 Orbits

To simplify the subsequent discussion, we shall refer to mass m as the *satellite* and to mass M as the *master*, keeping in mind that, in principle, we could also do the opposite.

Equation (11.18) can be treated like an equation for a Kepler orbit. For the *relative* orbit of the satellite about its master, we obtain

$$r = r_1 = \frac{p}{1 + \varepsilon \cos\theta} \tag{11.21}$$

The master mass M is located in the occupied focus of this orbit, which is shown as Orbit 1 in Figure 11.2. For Figure 11.2, $M = 4m$ and $\varepsilon = 0.5$ were arbitrarily chosen.

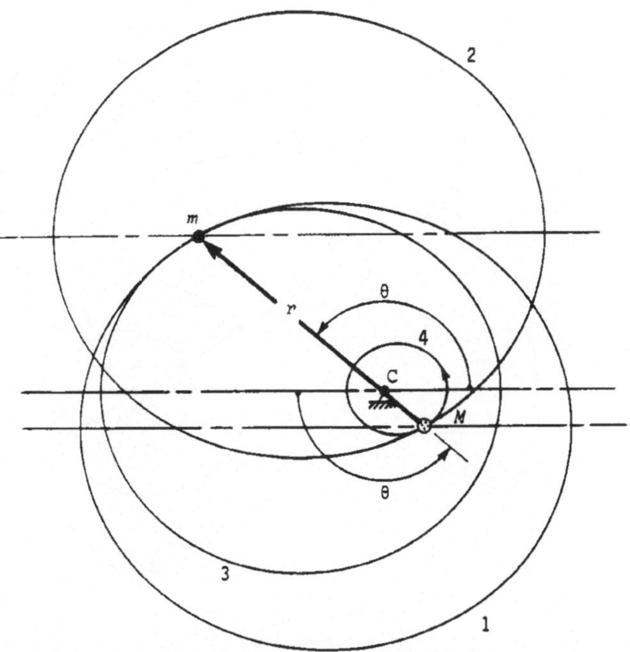

FIGURE 11.2. All four orbits have one and the same eccentricity.

If one now considers oneself located on m and determines the orbit of M about m, one obtains, for the *relative* orbit of the master about the satellite,

$$r_2 = \frac{p}{1 + \varepsilon \cos\theta} \qquad (11.22)$$

which is shown as Orbit 2 in Figure 11.2. (The angle θ is the true anomaly in equations (11.21) and (11.22), i.e. it is measured from the periapsis of each orbit.)

The *absolute* (i.e. with respect to the common mass centre C) orbit of satellite m is

$$r_3 = \frac{M}{M+m}r = \frac{p_3}{1 + \varepsilon \cos\theta} \qquad (11.23)$$

with the semi-parameter

$$p_3 = \frac{M}{M+m}p \qquad (11.24)$$

It appears as Orbit 3 in Figures 11.2 and 11.3.

For the *absolute* orbit of the master M, we have

$$r_4 = \frac{m}{M+m}r = \frac{p_4}{1 + \varepsilon \cos\theta} \qquad (11.25)$$

with the semi-parameter

$$p_4 = \frac{m}{M+m}p \qquad (11.26)$$

It appears as Orbit 4 in Figures 11.2 and 11.3. *All four orbits* shown in Figure 11.2, by the way, *have one and the same eccentricity*.

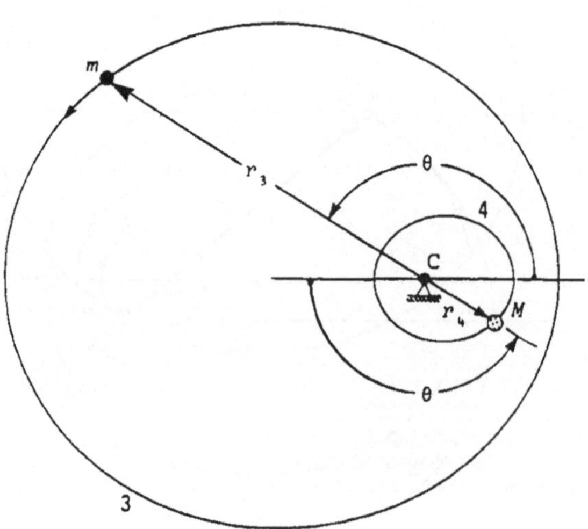

FIGURE 11.3. The absolute orbits.

11.2   The Earth-Moon System

The mass of the Earth, which we will consider as master, is

$$M = 597.6 \cdot 10^{22} \text{ kg}$$

The mass of the Moon is

$$m = 7.35 \cdot 10^{22} \text{ kg}$$

The *mass ratio* Moon/Earth is

$$\frac{m}{M} = 0.0123 \qquad (11.27)$$

or

$$\frac{m}{M} = \frac{1}{81.3} \qquad (11.28)$$

The gravitational parameter for the Moon-Earth system is

$$\mu' = \mu\left[1 + \frac{m}{M}\right] = 1.0123\,\mu \qquad (11.29)$$

where μ is the gravitational parameter of the Earth alone. With $\mu = 398\,601.19 \text{ km}^3/\text{s}^2$, we obtain

$$\mu' = 403\,504 \text{ km}^3/\text{s}^2$$

The eccentricity (Figure 11.4) of the absolute and relative elliptic orbits of the Earth-Moon system is

$$\varepsilon = 0.0549$$

The semi-major axis of the Moon's orbit about the Earth is approximately

$$a = 384\,400 \text{ km}$$

the semi-minor axis is approximately

$$b = 383\,365 \text{ km}$$

the semi-parameter is approximately

$$p = 383\,241 \text{ km}$$

11.3   The Sun - (Earth + Moon) System

The mass of the Sun, which we will choose as the master mass M, is

$$M = 1\,989 \cdot 10^{24} \text{ kg}$$

The mass m of (Earth + Moon) is

$$m = 6.0495 \cdot 10^{24} \text{ kg}$$

The mass ratio is consequently

$$\frac{m}{M} = \frac{1}{328\,455} = 0.000\,003\,04$$

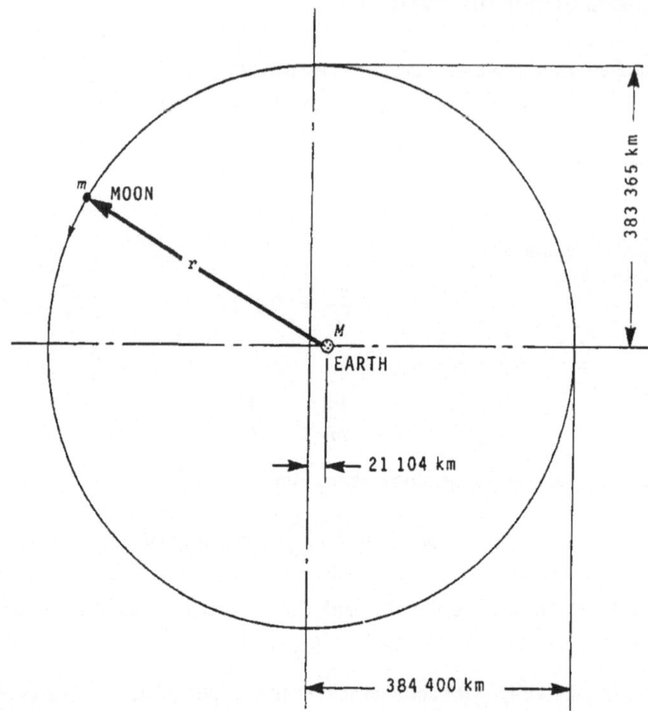

FIGURE 11.4. The ellipse of the Moon's orbit around the Earth.

The gravitational parameter of the system is

$$\mu' = \mu\left[1 + \frac{m}{M}\right] = 1.000\ 003\ 04\ \mu$$

where μ is the gravitational parameter of the Sun alone. With $\mu = 132.7 \cdot 10^9\ \text{km}^3/\text{s}^2$, we obtain

$$\mu' = 132.700\ 403 \cdot 10^9\ \text{km}^3/\text{s}^2$$

The reader is invited to compare the data for the Sun - (Earth + Moon) system and the Earth-satellite system with those of the Earth-Moon system.

Problem

11.1   In the two-body problem with point masses M and m, the semi-major axis of orbit "1" of m about M is a_1 and known. The Kepler period of the motion of m about M is thus

$$\tau_1 = 2\pi\sqrt{\frac{a_1^3}{\mu'}}$$

Considering that $a_4 = \dfrac{m}{M + m}a_1$, what is the Kepler period τ_4 (in terms of a_1 and μ') of mass M on orbit "4" about the common mass centre C?

12

Satellites in the Earth-Moon System

12.1 The Moon

The Earth-Moon system is unique, in that no other planet within the solar system has a trabant[*] of such impressive magnitude, the Pluto/Charon system excepted. The ratio Moon mass/Earth mass is 1:81.3, representing a comparatively large value. Earth plus Moon are therefore often looked upon as a double star. The Moon ranks fourteenth amongst the bodies of the solar system: only the nine planets, Jupiter's trabants Ganymede and Kallisto, Saturn's trabant Titan, and Neptune's trabant Triton have a larger mass than the Moon.

The Moon moves in *locked* rotation about the Earth, i.e. Earth dwellers only see one side of the Moon. Its orbit about the Earth takes about 27 days. Photographs of the backside of the Moon were obtained for the first time in October 1959 by the Soviet space probe "Luna 3". In January 1966, another unmanned Soviet space probe, "Luna 9", soft-landed successfully on the Moon, and on July 21, 1969, U.S. astronaut Neil A. Armstrong set foot on the Moon for the first time.

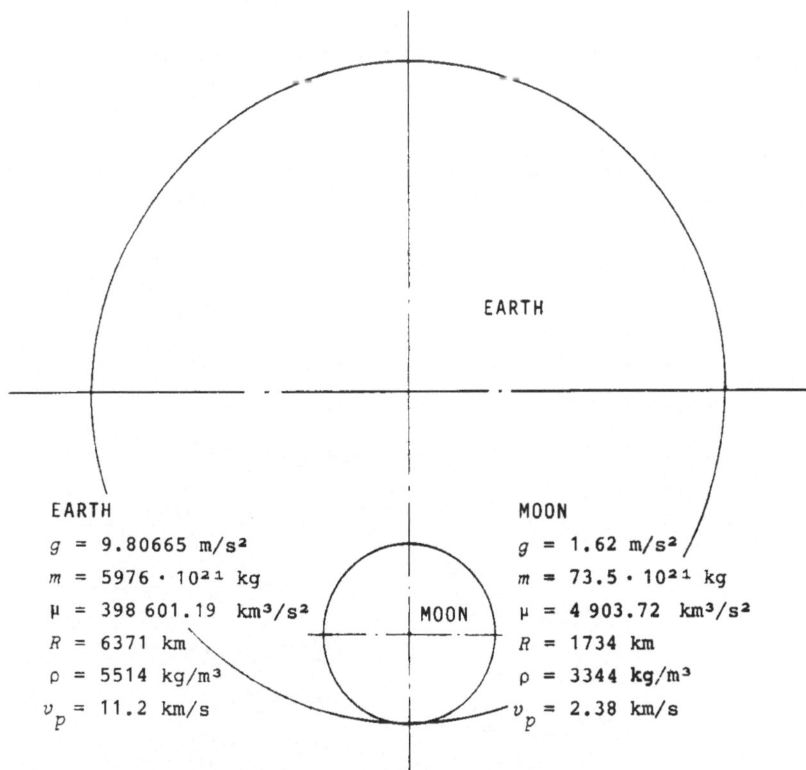

FIGURE 12.1. Comparison between Earth and Moon.

[*] The Czech term "trabant" is used here for moon-like bodies.

The mass of the Moon is roughly 1/81 of the Earth's mass. The Moon's diameter is roughly 1/4 of the Earth's diameter (Figure 12.1). The Moon's surface gravitational attraction constant is about 1/6 of the Earth's. The escape speed from the Moon's surface is about 1/5 of the escape speed from the Earth's surface. A Schuler period about the Moon is 1.28 times as large as that about the Earth.

The Moon's "day" (i.e. the period from high noon to high noon) lasts some 29 days. At a given location on the Moon's surface, there is thus some 14.5 days sunshine, and for 14.5 days it is night. During this night, the surface temperature on the Moon goes down to -150°C. During sunshine, it reaches +135°C. The high temperature and the low attraction force combined make it easy for gas molecules to escape. An atmosphere on the Moon cannot be retained.

In celestial mechanics, only the *two-body problem* has an unrestricted general solution. For the general *three-body problem* (as for the general *n-body problem* for that matter), there is no unrestricted general solution. If a spacecraft m, which is now a satellite of two masters, viz. the Earth M_E and the Moon M_M, has a mass that is negligible in comparison with each of the two master masses, then the motion of the two master masses is not influenced by the presence of the satellite, while the motion of the satellite, on the other hand, is fully dictated by the two master masses. This situation is typical for a man-made spacecraft in the Earth-Moon system. In celestial mechanics terminology, it represents a *restricted three-body problem.*

12.2   A Point Satellite in the Earth-Moon System

The Moon's orbit about the Earth has a very weak ellipticity. Figure 12.2 is intended to give the reader an idea of the geometry involved. We simplify the problem for the subsequent analysis, and assume that the Moon moves on a circular orbit, about the common mass centre 0 of Earth and Moon. The radius of the Earth's orbit about the common mass centre is $r_E = 4\,670\,\text{km}$. The radius of the Moon's orbit is $r_M = 379\,730\,\text{km}$ (Figure 12.3). The sum of the two represents the mean distance between Earth centre and Moon centre

$$d = r_E + r_M = 4\,670 + 379\,730 = 384\,400\,\text{km} \tag{12.1}$$

We shall assume in addition, that Earth as well as Moon are of spherical shape, and that the satellite's mass m is so small that its presence neither affects the location of the common mass centre 0, nor the orbits of Earth and Moon.

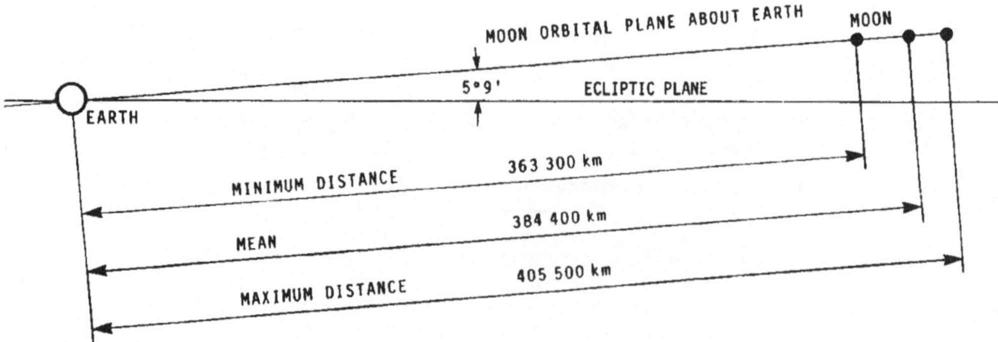

FIGURE 12.2. The Earth-Moon system.

Figure 12.3 shows the Earth-Moon system's mass centre 0 which serves as the origin of the coordinate system. The x-axis coincides with the line connecting Earth centre and Moon centre. From $\omega^2 = (\mu_E + \mu_M)/d^3$, with $\mu_E = 398\ 601.19\ \text{km}^3/\text{s}^2$, $\mu_M = 4902.73\ \text{km}^3/\text{s}^2$ and $d = 384\ 400$ km, this line rotates at an angular velocity of (Figure 12.4)

$$\omega = 2.665\ 317 \cdot 10^{-6}\ \text{rad/s} \tag{12.2}$$

The position vector $\mathbf{r}_E$ points from the system mass centre 0 to the mass centre of the Earth,

$$\mathbf{r}_E = -\frac{M_M}{M_E + M_M}\mathbf{d} \tag{12.3}$$

The position vector $\mathbf{r}_M$ points from the system mass centre 0 to the mass centre of the Moon,

$$\mathbf{r}_M = \frac{M_E}{M_E + M_M}\mathbf{d} \tag{12.4}$$

The vector $\mathbf{d}$, emanating at the centre of the Earth and locating the centre of the Moon, is consequently

$$\mathbf{d} = \mathbf{r}_M - \mathbf{r}_E \tag{12.5}$$

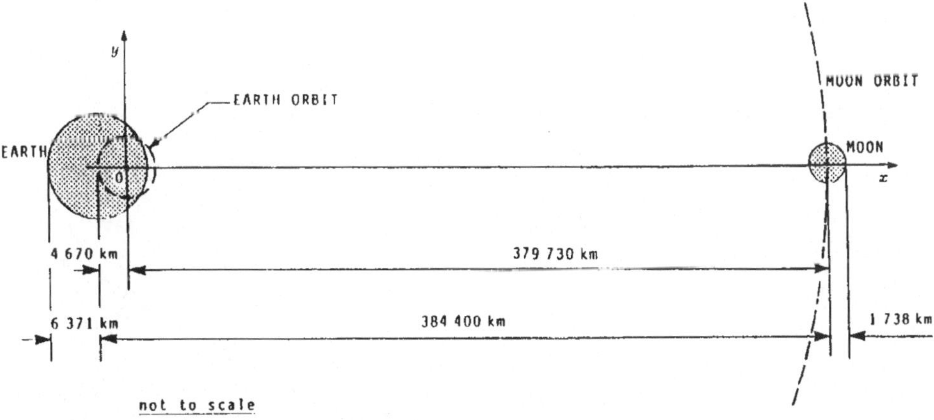

FIGURE 12.3. System mass centre 0 and circular orbits of Earth and Moon.

The information represented in planar form in Figure 12.3 is repeated, but this time in three dimensions, in Figure 12.5.

In Figure 12.5, the z-axis can be seen, which is perpendicular to the lunar orbital plane. The position vector $\mathbf{r}$ of the satellite m is

$$\mathbf{r} = \begin{bmatrix} \mathbf{e}_x\ \mathbf{e}_y\ \mathbf{e}_z \end{bmatrix} \begin{bmatrix} x \\ y \\ z \end{bmatrix} \tag{12.6}$$

The absolute velocity $\dot{\mathbf{r}}$ of the satellite is obtained by differentiation with respect to time,

$$\dot{\mathbf{r}} = \begin{bmatrix} \mathbf{e}_x\ \mathbf{e}_y\ \mathbf{e}_z \end{bmatrix} \begin{bmatrix} \dot{x} \\ \dot{y} \\ \dot{z} \end{bmatrix} + \begin{bmatrix} \dot{\mathbf{e}}_x\ \dot{\mathbf{e}}_y\ \dot{\mathbf{e}}_z \end{bmatrix} \begin{bmatrix} x \\ y \\ z \end{bmatrix} \tag{12.7}$$

or

$$\dot{\mathbf{r}} = \overset{\circ}{\mathbf{r}} + \boldsymbol{\omega} \times \mathbf{r} \tag{12.8}$$

where

$$\overset{\circ}{\mathbf{r}} \;=\; \Big[\, \mathbf{e}_x\, \mathbf{e}_y\, \mathbf{e}_z \,\Big] \begin{bmatrix} \dot{x} \\ \dot{y} \\ \dot{z} \end{bmatrix} \tag{12.9}$$

and

$$\boldsymbol{\omega}\times\mathbf{r} \;=\; \begin{vmatrix} \mathbf{e}_x & \mathbf{e}_y & \mathbf{e}_z \\ 0 & 0 & \omega \\ x & y & z \end{vmatrix} \;=\; \Big[\, \dot{\mathbf{e}}_x\, \dot{\mathbf{e}}_y\, \dot{\mathbf{e}}_z \,\Big] \begin{bmatrix} x \\ y \\ z \end{bmatrix} \tag{12.10}$$

The absolute acceleration is obtained by differentiating a second time. Recalling that ω = constant,

$$\ddot{\mathbf{r}} \;=\; (\overset{\circ}{\mathbf{r}}+\boldsymbol{\omega}\times\mathbf{r})^{o} \;+\; \boldsymbol{\omega}\times(\overset{\circ}{\mathbf{r}}+\boldsymbol{\omega}\times\mathbf{r}) \tag{12.11}$$

$$=\; \overset{\circ\circ}{\mathbf{r}} \;+\; \boldsymbol{\omega}\times\overset{\circ}{\mathbf{r}} \;+\; \boldsymbol{\omega}\times\overset{\circ}{\mathbf{r}} \;+\; \boldsymbol{\omega}\times(\boldsymbol{\omega}\times\mathbf{r}) \tag{12.12}$$

We expand the triple cross product

$$\boldsymbol{\omega}\times(\boldsymbol{\omega}\times\mathbf{r}) \;=\; (\boldsymbol{\omega}\cdot\mathbf{r})\boldsymbol{\omega} \;-\; (\boldsymbol{\omega}\cdot\boldsymbol{\omega})\mathbf{r} \;=\; (\boldsymbol{\omega}\cdot\mathbf{r})\boldsymbol{\omega} \;-\; \omega^2\mathbf{r} \tag{12.13}$$

and obtain for the absolute acceleration

$$\ddot{\mathbf{r}} \;=\; \overset{\circ\circ}{\mathbf{r}} \;+\; 2\boldsymbol{\omega}\times\overset{\circ}{\mathbf{r}} \;+\; (\boldsymbol{\omega}\cdot\mathbf{r})\boldsymbol{\omega} \;-\; \omega^2\mathbf{r} \tag{12.14}$$

The vector $(\boldsymbol{\omega}\cdot\mathbf{r})\boldsymbol{\omega}$ points in the z-direction since $\boldsymbol{\omega}$ itself points that way.

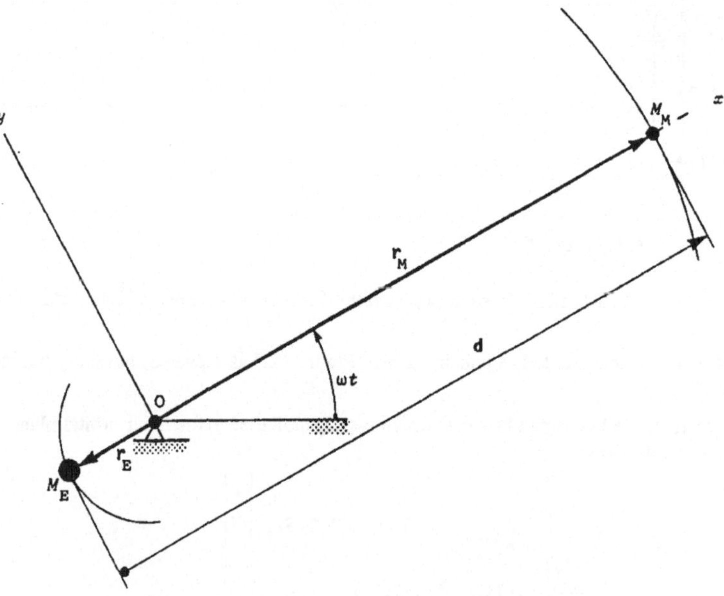

FIGURE 12.4. Rotation of the Earth-Moon system.

We can thus write

$$(\boldsymbol{\omega}\cdot\mathbf{r})\boldsymbol{\omega} \;=\; (\omega z)\boldsymbol{\omega} \;=\; \omega^2 z\,\mathbf{e}_z \tag{12.15}$$

Eventually, one arrives at an expression for the absolute acceleration of the satellite which has the following form

$$\ddot{\mathbf{r}} = \overset{\circ\circ}{\mathbf{r}} + 2\boldsymbol{\omega}\times\overset{\circ}{\mathbf{r}} - \omega^2(\mathbf{r} - z\mathbf{e}_z) \tag{12.16}$$

According to Newton's second and inverse square laws, we can write for a small point satellite of mass m

$$m\ddot{\mathbf{r}} = -\frac{\mu_E m}{d_E^3}(\mathbf{r} - \mathbf{r}_E) - \frac{\mu_M m}{d_M^3}(\mathbf{r} - \mathbf{r}_M) \tag{12.17}$$

where

$$|\mathbf{r} - \mathbf{r}_E| = d_E \quad \text{and} \quad |\mathbf{r} - \mathbf{r}_M| = d_M$$

If equations (12.16) and (12.17) are combined, one can write

$$\overset{\circ\circ}{\mathbf{r}} + 2\boldsymbol{\omega}\times\overset{\circ}{\mathbf{r}} - \omega^2(\mathbf{r} - z\mathbf{e}_z) = -\frac{\mu_E}{d_E^3}(\mathbf{r} - \mathbf{r}_E) - \frac{\mu_M}{d_M^3}(\mathbf{r} - \mathbf{r}_M) \tag{12.18}$$

Since however

$$\omega^2(\mathbf{r} - z\mathbf{e}_z) = \omega^2(x\mathbf{e}_x + y\mathbf{e}_y) = \mathbf{grad}\frac{\omega^2}{2}(x^2 + y^2) \tag{12.19}$$

and furthermore

$$-\frac{\mu_E}{d_E^3}(\mathbf{r} - \mathbf{r}_E) = \mathbf{grad}\frac{\mu_E}{d_E} \tag{12.20}$$

and

$$-\frac{\mu_M}{d_M^3}(\mathbf{r} - \mathbf{r}_M) = \mathbf{grad}\frac{\mu_M}{d_M} \tag{12.21}$$

equation (12.18) can also be written

$$\overset{\circ\circ}{\mathbf{r}} + 2\boldsymbol{\omega}\times\overset{\circ}{\mathbf{r}} = \mathbf{grad}\left[\frac{\mu_E}{d_E} + \frac{\mu_M}{d_M} + \frac{\omega^2}{2}(x^2 + y^2)\right] \tag{12.22}$$

Dot multiplication with $\overset{\circ}{\mathbf{r}}$ gives

$$\overset{\circ}{\mathbf{r}}\cdot\overset{\circ\circ}{\mathbf{r}} = \overset{\circ}{\mathbf{r}}\cdot\mathbf{grad}\left[\frac{\mu_E}{d_E} + \frac{\mu_M}{d_M} + \frac{\omega^2}{2}(x^2 + y^2)\right] \tag{12.23}$$

The following relationship applies to any arbitrary differentiable scalar function U

$$\overset{\circ}{\mathbf{r}}\cdot\mathbf{grad}U = \begin{bmatrix}\mathbf{e}_x & \mathbf{e}_y & \mathbf{e}_z\end{bmatrix}\begin{bmatrix}\dot{x}\\\dot{y}\\\dot{z}\end{bmatrix}\cdot\begin{bmatrix}\mathbf{e}_x & \mathbf{e}_y & \mathbf{e}_z\end{bmatrix}\begin{bmatrix}\dfrac{\partial U}{\partial x}\\[2mm]\dfrac{\partial U}{\partial y}\\[2mm]\dfrac{\partial U}{\partial z}\end{bmatrix} \tag{12.24}$$

$$= \frac{\partial U}{\partial x}\frac{dx}{dt} + \frac{\partial U}{\partial y}\frac{dy}{dt} + \frac{\partial U}{\partial z}\frac{dz}{dt} = \frac{dU}{dt}$$

When applied to the case of equation (12.23), relation (12.24) means that

$$\overset{\circ}{\mathbf{r}}\cdot\overset{\circ\circ}{\mathbf{r}} = \frac{d}{dt}\left[\frac{\mu_E}{d_E} + \frac{\mu_M}{d_M} + \frac{\omega^2}{2}(x^2 + y^2)\right] \tag{12.25}$$

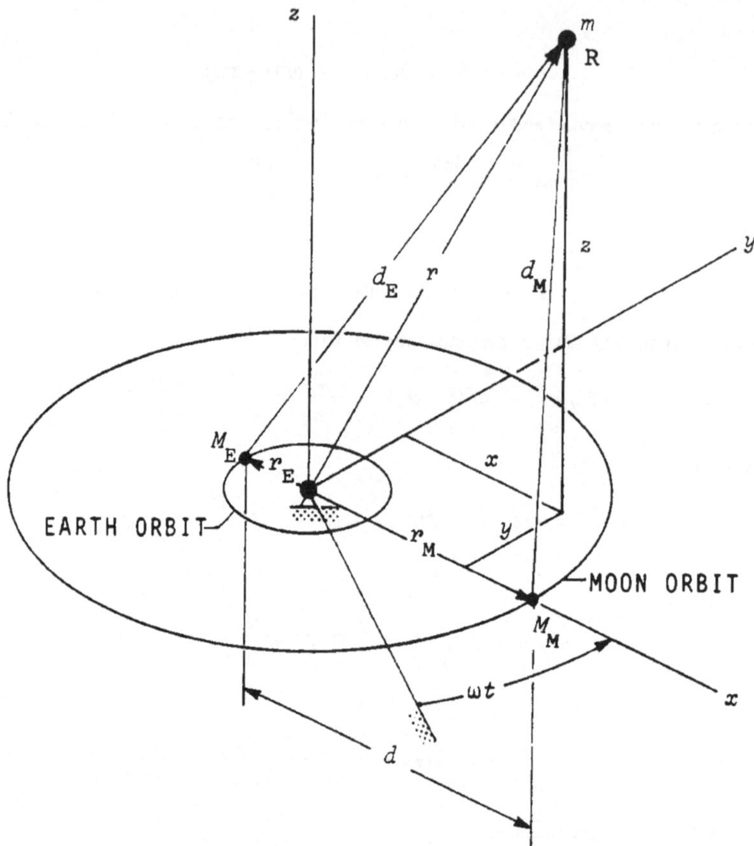

FIGURE 12.5. Satellite m in the Earth-Moon system.

Furthermore,

$$\frac{d}{dt}(\overset{\circ}{\mathbf{r}} \cdot \overset{\circ}{\mathbf{r}}) = \overset{\circ}{\overset{\cdot}{\mathbf{r}}} \cdot \overset{\circ}{\mathbf{r}} + \overset{\circ}{\mathbf{r}} \cdot \overset{\circ}{\overset{\cdot}{\mathbf{r}}}$$

$$= 2\,\overset{\circ}{\mathbf{r}} \cdot \overset{\circ}{\overset{\cdot}{\mathbf{r}}}$$

$$= 2\,\overset{\circ}{\mathbf{r}} \cdot (\overset{\circ\circ}{\mathbf{r}} + \boldsymbol{\omega} \times \overset{\circ}{\mathbf{r}}) \tag{12.26}$$

$$\frac{d}{dt}\frac{1}{2}(\overset{\circ}{\mathbf{r}} \cdot \overset{\circ}{\mathbf{r}}) = \overset{\circ}{\mathbf{r}} \cdot \overset{\circ\circ}{\mathbf{r}} + \overset{\circ}{\mathbf{r}} \cdot (\boldsymbol{\omega} \times \overset{\circ}{\mathbf{r}})$$

$$= \overset{\circ}{\mathbf{r}} \cdot \overset{\circ\circ}{\mathbf{r}}$$

With the aid of equation (12.26), equation (12.25) becomes

$$\frac{d}{dt}\frac{1}{2}(\overset{\circ}{\mathbf{r}} \cdot \overset{\circ}{\mathbf{r}}) = \frac{d}{dt}\left[\frac{\mu_E}{d_E} + \frac{\mu_M}{d_M} + \frac{\omega^2}{2}(x^2 + y^2)\right] \tag{12.27}$$

On the other hand, we may write

$$\overset{\circ}{\mathbf{r}} \cdot \overset{\circ}{\mathbf{r}} = \dot{x}^2 + \dot{y}^2 + \dot{z}^2 = v^2 \tag{12.28}$$

where v is the *relative* velocity of the satellite m with respect to point R (Figure 12.5), which is fixed within the rotating Earth-Moon system. Thus, we may write

$$\frac{d}{dt}\left[\frac{v^2}{2}\right] = \frac{d}{dt}\left[\frac{\mu_E}{d_E} + \frac{\mu_M}{d_M} + \frac{\omega^2}{2}(x^2 + y^2)\right] \tag{12.29}$$

or upon integration (Jacobi[*] integral)

$$v^2 = 2\frac{\mu_E}{d_E} + 2\frac{\mu_M}{d_M} + \omega^2(x^2 + y^2) - C \tag{12.30}$$

where C is the integration constant. Its value is a characteristic of any given satellite and can be obtained from equation (12.30), provided all other quantities are known for the instant under consideration.

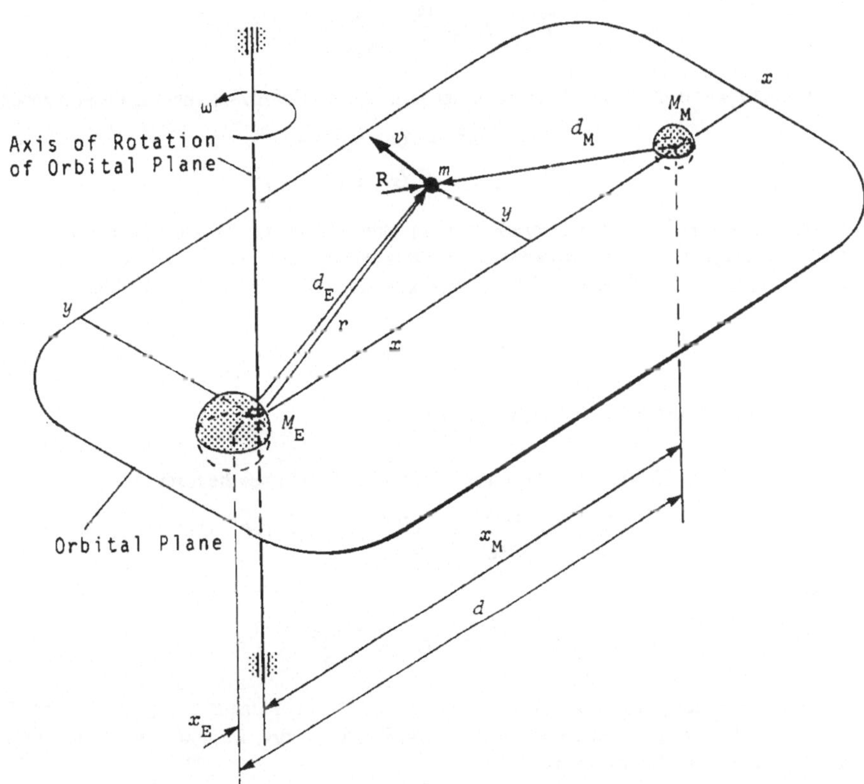

FIGURE 12.6. Satellite m within orbital plane.

12.3 Satellites Within the Lunar Orbital Plane

Let us now consider a case of practical importance, the satellite which moves within the plane of the Moon's orbit about the Earth. The satellite is located within the orbital plane, i.e. $z = 0$, and its velocity is also within the orbital plane, i.e. $v_z = 0$.

We employ equation (12.23), which states that

$$\mathring{\mathbf{r}} \cdot \left\{ \mathring{\mathring{\mathbf{r}}} - \text{grad} \left[\frac{\mu_E}{d_E} + \frac{\mu_M}{d_M} + \frac{\omega^2}{2}(x^2 + y^2) \right] \right\} = 0 \tag{12.31}$$

---

[*] Karl Gustav Jacob Jacobi (1804-1851), German mathematician

and conclude that the vector

$$\overset{\circ\circ}{r} \; - \; \text{grad}\left[\frac{\mu_E}{d_E} + \frac{\mu_M}{d_M} + \frac{\omega^2}{2}(x^2 + y^2)\right] \tag{12.32}$$

must be perpendicular to $\overset{\circ}{r}$. Furthermore, equation (12.22) demands that

$$\overset{\circ\circ}{r} \; - \; \text{grad}\left[\frac{\mu_E}{d_E} + \frac{\mu_M}{d_M} + \frac{\omega^2}{2}(x^2 + y^2)\right] \; = \; -2\omega \times \overset{\circ}{r} \tag{12.33}$$

Thus, if $\overset{\circ}{r}$ lies within the xy-plane, and if ω is perpendicular to that plane, then

$$\overset{\circ\circ}{r} \; - \; \text{grad}\left[\frac{\mu_E}{d_E} + \frac{\mu_M}{d_M} + \frac{\omega^2}{2}(x^2 + y^2)\right] \tag{12.34}$$

must also lie within the xy-plane. If the initial position vector of the satellite has no z-component, then

$$d_E^2 \; = \; (x - x_E)^2 \; + \; y^2$$
$$d_M^2 \; = \; (x - x_M)^2 \; + \; y^2 \tag{12.35}$$

i.e. the expression in the square brackets of equation (12.34) has no z-component either, and the same applies to the gradient of the expression. Thus it is shown that if neither the position nor the velocity have a z-component, the z-component of the relative acceleration $\overset{\circ\circ}{r}$ vanishes. The satellite will remain forever in the xy-plane.

12.4 An Energy Consideration

Consider equation (12.30), which by multiplying with $m/2$, assumes the form

$$\frac{mv^2}{2} \; = \; \frac{\mu_E m}{d_E} \; + \; \frac{\mu_M m}{d_M} \; + \; \frac{m\omega^2}{2}(x^2 + y^2) \; + \; C_1 \tag{12.36}$$

where

$$C_1 \; = \; -\frac{Cm}{2}$$

The left hand side of equation (12.36) represents some form of kinetic energy. Unfortunately, v is merely a relative velocity. In order to obtain the actual kinetic energy, we note that $v^2 \; = \; |\overset{\circ}{r}|^2$, and that the absolute velocity of the satellite is

$$\dot{r} \; = \; \overset{\circ}{r} \; + \; \omega \times r \tag{12.37}$$

such that equation (12.36) becomes

$$\frac{m}{2}\left|\dot{r} - \omega \times r\right|^2 \; = \; \frac{\mu_E m}{d_E} \; + \; \frac{\mu_M m}{d_M} \; + \; \frac{m\omega^2}{2}(x^2 + y^2) \; + \; C_1 \tag{12.38}$$

Instead of $\omega \times r$, one may write

$$\omega \times r \; = \; \begin{vmatrix} e_x & e_y & e_z \\ 0 & 0 & \omega \\ x & y & z \end{vmatrix} \; = \; \omega(xe_y - ye_x)$$

and thus

$$|\omega \times r|^2 \; = \; \omega^2(x^2 + y^2) \tag{12.39}$$

such that equation (12.38) is simplified to

$$\frac{m}{2}\left\{ |\dot{\mathbf{r}}|^2 - 2\dot{\mathbf{r}}\cdot(\boldsymbol{\omega}\times\mathbf{r}) \right\} = \frac{\mu_E m}{d_E} + \frac{\mu_M m}{d_M} + C_1 \tag{12.40}$$

Finally, we introduce

$$\boldsymbol{\omega}\times\mathbf{r} = \mathbf{v}_R \tag{12.41}$$

where $\mathbf{v}_R$ is the absolute velocity of the point R which is fixed in the Earth-Moon system and happens to be located at a position $\mathbf{r}$, which the satellite just passes at an absolute velocity $\dot{\mathbf{r}}$.

Now we can write for the total energy of the satellite

$$E = \frac{m|\dot{\mathbf{r}}|^2}{2} - \frac{\mu_E m}{d_E} - \frac{\mu_M m}{d_M} = C_1 + m\dot{\mathbf{r}}\cdot\mathbf{v}_R \tag{12.42}$$

The term $\dfrac{m|\dot{\mathbf{r}}|^2}{2}$ is the kinetic energy of the satellite, while $-\dfrac{\mu_E m}{d_E} - \dfrac{\mu_M m}{d_M}$ represents the satellite's potential energy.

The total energy of the Earth-Moon system (which we have assumed to be isolated) is constant. Equation (12.42) thus informs us that the satellite continually withdraws energy from, or returns energy to, the Earth-Moon system.

12.5 Hill's Curves

The relative velocity v of a satellite in the Earth-Moon system is given by

$$v^2 = 2\frac{\mu_E}{d_E} + 2\frac{\mu_M}{d_M} + \omega^2(x^2 + y^2) - C \tag{12.43}$$

where

$$d_E^2 = (x - x_E)^2 + y^2 + z^2$$

and

$$d_M^2 = (x - x_M)^2 + y^2 + z^2$$

with $z = 0$ for satellites within the Moon's orbital plane about the Earth. The quantity ω is the angular velocity of the x-axis, which coincides with the line connecting Earth centre and Moon centre. The z-axis is the axis of rotation.

We may calculate the magnitude of the constant C for a given satellite m if the relative speed v_i of the satellite, its distance d_{Ei} from Earth centre and its distance d_{Mi} from Moon centre, and its instantaneous position coordinates x_i and y_i are known for a given specific instant i.

$$C = 2\frac{\mu_E}{d_{Ei}} + 2\frac{\mu_M}{d_{Mi}} + \omega^2(x_i^2 + y_i^2) - v_i^2 \tag{12.44}$$

Spacecraft on lunar missions are generally inserted into an orbit that lies within the orbital plane of the Moon's motion about the Earth. Orbits outside the orbital plane require more fuel and are thus only used when the mission profile specifically prescribes it.

The influence of the Sun can also normally be neglected for Moon missions, since a typical Moon mission lasts for a few days only.

We thus can make the following assumptions:

1. Earth and Moon move in a plane fixed in space.

2. The satellite moves within the same plane.

3. The satellite's mass is so small that its influence on the Earth and Moon orbits is negligible.

4.   Earth and Moon both move on circular orbits.

5.   The ratio of Moon mass M_M to Earth mass M_E is

$$\frac{M_M}{M_E} = \frac{\mu_M}{\mu_E} = \frac{4\,902.73}{398\,601.19} = 0.0123 = \frac{1}{81.3} \tag{12.45}$$

6.   the influence of other celestial bodies, such as the Sun, is negligible.

Considering the relative speed of a satellite in the Earth-Moon system as given by equation (12.43), one concludes that the speed changes as the position of the satellite in the Earth-Moon system changes. We can now try and determine those positions of the satellite for which the speed vanishes. The speed is zero if

$$2\frac{\mu_E}{d_E} + 2\frac{\mu_M}{d_M} + \omega^2(x^2+y^2) = C \tag{12.46}$$

For satellites within the orbital plane (Figure 12.6), the distances d_E and d_M are obtainable from

$$d_E^2 = (x-x_E)^2 + y^2$$
$$d_M^2 = (x-x_M)^2 + y^2$$

Thus

$$2\frac{\mu_E}{\sqrt{(x-x_E)^2+y^2}} + 2\frac{\mu_M}{\sqrt{(x-x_M)^2+y^2}} + \omega^2(x^2+y^2) = C \tag{12.47}$$

becomes a curve $y = f(x)$ in the Moon's orbital plane.

Equation (12.47) represents a curve for a given constant C within the orbital plane, known as *Hill's curve*[*] (or *Darwin's curve*[**]). Hill's curve represents a limit beyond which the satellite (or a stone thrown from the surface of the Earth), whose characteristic constant is C, cannot move. If the Hill curve is reached by a satellite, then its relative speed is zero, not so, however, its acceleration, under normal circumstances. Thus the satellite will return from its Hill curve, and will gather speed again.

For other values of C, there are other Hill curves. The whole family of Hill curves for the Earth-Moon system is shown in Figure 12.7.

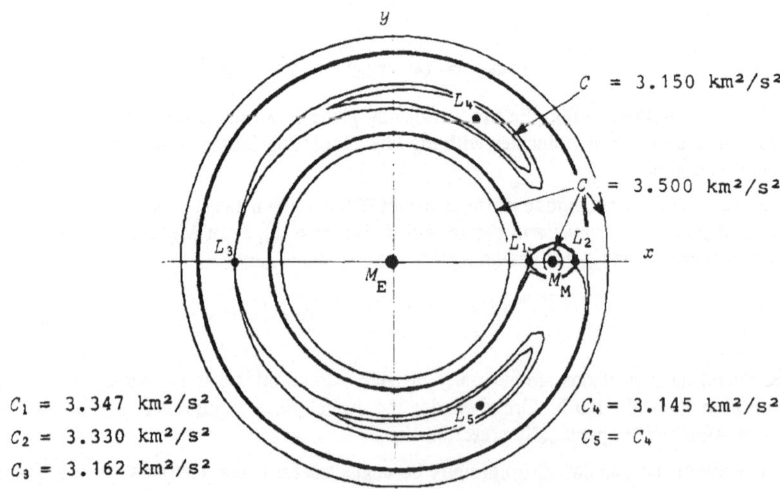

FIGURE 12.7. Hill's curves in the Earth-Moon orbital plane.

---

[*] George William Hill (1838-1914), American astronomer

[**] George Howard Darwin (1845-1912), English physicist

Of special interest are the singularities L_1, L_2, L_3, L_4 and L_5. They are the Lagrangian[*] *libration points*, and represent points at which both the *relative velocity* and the *relative acceleration* of the satellite are zero.

12.6 Libration Points

Points at which both, the relative velocity $\overset{\circ}{r}$ and the relative acceleration $\overset{\circ\circ}{r}$ are zero, are called *libration points*. Returning to equation (12.22), results in

$$\mathbf{grad}\left[\frac{\mu_E}{d_E} + \frac{\mu_M}{d_M} + \frac{\omega^2}{2}(x^2 + y^2)\right] = 0 \tag{12.48}$$

We now restrict the problem to positions within the lunar orbital plane by setting $z = 0$, such that

$$d_E = \sqrt{(x - x_E)^2 + y^2} \tag{12.49}$$

and

$$d_M = \sqrt{(x - x_M)^2 + y^2} \tag{12.50}$$

The x-component of equation (12.48) is

$$\frac{\partial}{\partial x}\left[\frac{\mu_E}{\sqrt{(x - x_E)^2 + y^2}} + \frac{\mu_M}{\sqrt{(x - x_M)^2 + y^2}} + \frac{\omega^2}{2}(x^2 + y^2)\right] = 0 \tag{12.51}$$

giving

$$-\frac{\mu_E(x - x_E)}{\sqrt{(x - x_E)^2 + y^2}\,^3} - \frac{\mu_M(x - x_M)}{\sqrt{(x - x_M)^2 + y^2}\,^3} + \omega^2 x - 0 \tag{12.52}$$

or

$$\omega^2 x\sqrt{(x - x_E)^2 + y^2}\,^3\,\sqrt{(x - x_M)^2 + y^2}\,^3 - \mu_E(x - x_E)\sqrt{(x - x_M)^2 + y^2}\,^3 \tag{12.53}$$
$$- \mu_M(x - x_M)\sqrt{(x - x_E)^2 + y^2}\,^3 = 0$$

By a similar process, we obtain for the y-component

$$\omega^2 y\sqrt{(x - x_E)^2 + y^2}\,^3\,\sqrt{(x - x_M)^2 + y^2}\,^3 - \mu_E y\sqrt{(x - x_M)^2 + y^2}\,^3$$
$$- \mu_M y\sqrt{(x - x_E)^2 + y^2}\,^3 = 0 \tag{12.54}$$

From equation (12.53) and (12.54), the five libration points can be obtained (Table 12.1)

Libration Point	From Earth d_E	From Moon d_M
L_1	$0.85\,d$	$0.15\,d$
L_2	$1.17\,d$	$0.17\,d$
L_3	$0.99\,d$	$1.99\,d$
L_4	$1.00\,d$	$1.00\,d$
L_5	$1.00\,d$	$1.00\,d$

d = distance between Earth centre and Moon centre = 384 400 km

TABLE 12.1: Location of libration points.

---

[*] Joseph Louis Lagrange (1736-1813), French mathematician.

One finds two libration points, L_4 and L_5, located such that the points M_E, M_M and L_4 (or L_5) form an equilateral triangle (Figure 12.8). The other points lie on the x-axis and satisfy equation (12.53) in the following form

$$\omega^2 x\,|x - x_E|^3\,|x - x_M|^3 \;-\; \mu_E(x - x_E)|x - x_M|^3 \;-\; \mu_M(x - x_M)|x - x_E|^3 \;=\; 0$$

giving

$$\xi_1 \;=\; x_1 \;-\; x_E \;=\; 326\,740\ \text{km}$$

$$\xi_2 \;=\; x_2 \;-\; x_E \;=\; 449\,784\ \text{km}$$

$$\xi_3 \;=\; x_3 \;-\; x_E \;=\; -380\,556\ \text{km}$$

with $x_E \;=\; -4670$ km, and the ξ's measured from Earth centre.

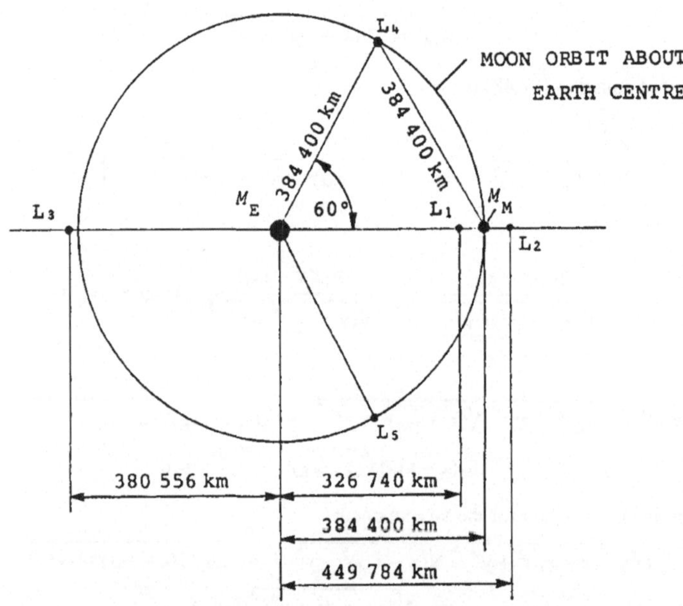

FIGURE 12.8. Libration points of Earth-Moon system.

It can be shown (Leipholz, 1987), and an inspection of Figure 12.7 suggests that within the theoretical treatment presented here, a satellite which happens to rest in libration point L_4 or L_5 is stable, i.e. will remain at the point and move along permanently with the Earth-Moon system, while a satellite located in libration point L_1, L_2, or L_3 is unstable, i.e. will wander off if slightly disturbed. Because of the simplifying assumptions made, several minor effects have been neglected, which will make their presence felt on a satellite resting in one of the libration points. Thus some form of active station-keeping is required to keep a satellite at a libration point, be it a stable or an unstable one.

Suggested Reading

1.   Leipholz, H. *Stability Theory* Second Edition, John Wiley, 1987, 359 pp.

Problems

12.1 Given is the equation

$$\omega^2 x \,|x - x_E|^3\,|x - x_M|^3 \quad \mu_E(x - x_E)|x - x_M|^3 \;-\; \mu_M(x - x_M)|x - x_E|^3 \;=\; 0$$

Determine the three real roots (in addition to $x = x_E$ and $x = x_M$) of x, for

$$\omega = 2.665\,317 \cdot 10^{-6}\ \text{rad/s}$$

$$x_E = -4\,670\ \text{km}$$

$$x_M = 379\,730\ \text{km}$$

$$\mu_E = 398\,601.19\ \text{km}^3/\text{s}^2$$

$$\mu_M = 4\,902.73\ \text{km}^3/\text{s}^2$$

12.2 An unpowered spacecraft m happens to be located on the line that connects Earth centre and Moon centre. The spacecraft is thus subjected to the Earth's attraction, and Moon's attraction, and a centrifugal force due to the line's rotation. At which distance from Earth centre would the spacecraft have to be located such that the sum of the three forces acting on it is zero?

12.3 Due to friction losses brought about by the tidal activity of its oceans, the Earth's angular momentum is continually reduced and transferred to the Moon, such that the combined angular momentum remains constant. Show that if the radius of the Moon's orbit increases by 4 cm each year, the Earth's day becomes longer by some 2 ms in a century. Make simplifying assumptions, such as that the Moon's orbit is within the Earth's equatorial plane, that the Moon's orbit is circular (384 400 km), that the Moon is a point mass, and that there is no perturbation from the Sun. The common mass centre is 4670 km from the Earth's centre. Also show that the Earth-Moon system loses kinetic energy.

12.4 Determine the Jacobian constant C for a satellite remaining at libration point L_4 in the Earth-Moon system.

12.5 The Trojans are planetoids located in libration points L_4 and L_5 of the Sun-Jupiter system. (a) What is the value of teir Jacobian constant C? (b) How far away from the Sun are they located? (c) What is their absolute speed?

13

Tidal Forces

One of the basic concerns in spacecraft engineering is to establish whether loose parts (e.g. tools) on the outside of a spacecraft tend to remain with the spacecraft or to drift away.

Consider a satellite m_s of spherical shape with a radius a (Figure 13.1), which is moving on a circular orbit about its master M. Outside the satellite, there is a small point mass m, which is screwed to the satellite by means of a bolt. The distance between mass point m and satellite centre is a. The satellite m_s is rotating in locked rotation, that is in a manner such that the same side is always pointing towards the master. We shall further assume that $m \ll m_s \ll M$ and that $a \ll r$.

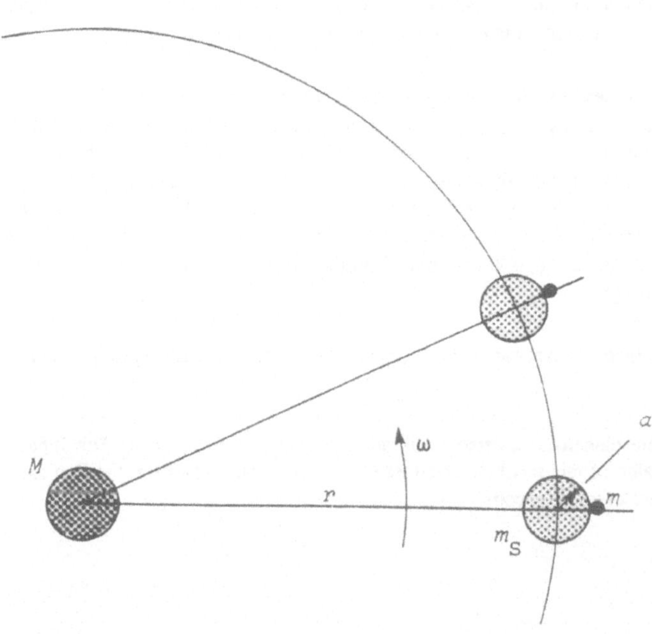

FIGURE 13.1. Master mass M, satellite mass m_s and point-mass m during locked rotation.

The small point mass m is subjected to the following forces:

F = force transmitted through bolt

K_M = Kepler force of primary master M

K_S = Kepler force of secondary master m_s

C = centrifugal force

such that (Figure 13.2)

$$F + K_M + K_S - C = 0 \tag{13.1}$$

For the attraction by the primary master, we have

$$K_M = \frac{GMm}{(r + a)^2}$$

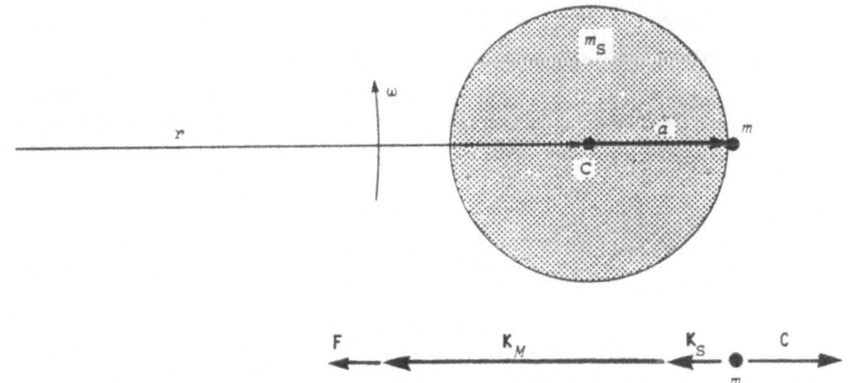

FIGURE 13.2. Forces acting on point mass m.

or upon expansion into a series, and breaking off after the second term,

$$K_M = \frac{GMm}{r^2}\left[1 - 2\frac{a}{r}\right] \tag{13.2}$$

The attraction by the satellite is

$$K_S = \frac{Gm_s m}{a^2} \tag{13.3}$$

The centrifugal force is

$$C = m\omega^2(r + a) \tag{13.4}$$

For the orbital angular velocity of the satellite, we have

$$\omega^2 = \frac{GM}{r^3}$$

such that the centrifugal force can be written

$$C = \frac{GMm}{r^2}\left[1 + \frac{a}{r}\right] \tag{13.5}$$

For the tension force transmitted through the bolt, one obtains eventually

$$F = 3\frac{GMma}{r^3} - \frac{Gm_s m}{a^2} \tag{13.6}$$

If the second term on the right hand side of equation (13.6) is greater than the first term, the force F becomes a compression force. The small mass m (e.g. a tool) will then remain with the satellite, even without a bolt connection.

In order to ensure that tools, astronauts, etc. will not drift away from a satellite, we thus have as a condition

$$3\frac{Ma}{r^3} < \frac{m_s}{a^2} \tag{13.7}$$

or

$$r^3 > 3\frac{M}{m_s}a^3 \tag{13.8}$$

where

M = primary master mass

m_s = mass of satellite

a = distance between mass m (e.g. tool) and mass centre of satellite

Relation (13.8) states in effect, that loose objects will drift away from *low-flying* satellites. This is an undesirable yet inevitable characteristic of significance for extra-vehicular activities (EVAs). For satellites far removed from their master, the problem does not arise. Loose subjects will then remain in contact with the satellite.

The quantity

$$r_R = \left[3\frac{M}{m_s} \right]^{1/3} a \tag{13.9}$$

is the *Roche limit**. One can take equation (13.9) a step further, and assume *spherical* mass distributions for satellite and primary master. Then

$$m_s = \frac{4}{3}\pi a^3 \rho_s \tag{13.10}$$

and

$$M = \frac{4}{3}\pi R^3 \rho_M \tag{13.11}$$

where ρ_s and ρ_M are the respective densities. If the expressions for the spherical masses are entered into equation (13.9), the Roche limit for a rigid spherical satellite becomes

$$r_R = 1.442 \left[\frac{\rho_M}{\rho_s} \right]^{1/3} R \tag{13.12}$$

For obvious reasons, artificial satellites are made of very light material. Furthermore, there may be a lot of empty space inside a satellite. Thus ρ_s is usually quite small.

The spherical balloon satellite Echo I ($\uparrow$1960-August-12, $\downarrow$1968-May-24) had a mass of 75 kg and radius of 15 m. This gave Echo I an average density of

$$\rho_s = \frac{3m_s}{4\pi a^3} \approx 0.005 \text{ kg/m}^3$$

while the Earth's density is roughly

$$\rho_M = 5514 \text{ kg/m}^3$$

such that

$$\frac{\rho_M}{\rho_s} \approx 10^6$$

Echo represents a satellite of extremely low density. More typical are probably satellite densities of, say,

$$\rho_s = 30 \text{ kg/m}^3$$

The Roche limit for a spherical satellite of this density circling the Earth would then be

$$r_R = 1.442 \left[\frac{5514}{30} \right]^{1/3} 6371 \approx 50\,000 \text{ km}$$

In other words, typical artificial satellites must have distances of more than 50 000 km from Earth centre before one can expect that loose tools, nuts and bolts, etc. and astronauts on extra-vehicular activities, tend

---

* Edouard Roche (1820-1883), French mathematician

to remain with the satellite.

A good example of a spherical satellite beyond the Roche limit is the Moon. For the Moon (ρ_s = 3344 kg/m^3) as a satellite of the Earth (ρ_M = 5514 kg/m^3), the Roche limit is at

$$r_R = 1.442 \left[\frac{5514}{3344} \right]^{1/3} 6371 = 10\ 858 \text{ km}$$

i.e. should the Moon ever come so close that it circles the Earth at less than 10 858 km, loose rocks will drift away from the Moon and the Moon will begin to disintegrate.

There exist also satellites with very high densities. The French "Starlette" class geodetic satellites are essentially spherical, have a radius of 120 mm and a mass of 47 kg (solid thorium core) such that their density amounts to some 6493 kg/m^3. The U.S. "Lageos" class satellites have 300 mm radius and a mass of 480 kg (solid brass core), giving them a density of some 4244 kg/m^3. For a Starlette on a circular orbit about the Earth, the Roche limit would be at

$$r_R = 1.442 \left[\frac{5514}{6493} \right]^{1/3} 6371 = 8701 \text{ km}$$

Based on rather complicated theory, the shapes of liquids subjected to eigengravitation and to centrifugal forces can be computed. If the secondary master m_s is such a liquid, rather than the solid assumed in the derivation above, it can be shown (Bucerius and Schneider, 1967) that that liquid would assume the shape of an ellipsoid. For this special case (with is e.g. of interest to Saturn's rings and trabants), the Roche limit becomes

$$r_R = 2.4554 \left[\frac{\rho_M}{\rho_s} \right]^{1/3} R \tag{13.13}$$

which is the value originally (1848) given by Roche.

Suggested Reading

1.   Bucerius, H.; M. Schneider, *Himmelsmechanik II*, B.I. Mannheim, 114/144a, 1967, 262 pp.

Problems

13.1   A satellite of spherical shape and 2 m radius, circles the Earth sphere at 1000 km above the surface in locked rotation. The uniformly distributed mass of the satellite is 500 kg. (a) What force is required to secure a 4 kg tool to the satellite's skin? (b) Where is the Roche limit?

13.2   A spherical spacecraft, of 3 m radius, moves in locked rotation about the Earth. During an EVA (Extra-Vehicular Activity) a small tool is placed by an astroworker on the far side of the spacecraft. The tool begins to drift away. Neglect the gravitational attraction between tool and spacecraft, and determine (a) the distance between tool and spacecraft when the latter has completed one half orbit about the Earth, and (b) how much time the astroworker has available to retrieve the tool before it drifts out of reach (1.2 m). The spacecraft is on a circular orbit of 7000 km radius about the Earth.

13.3   Sketch to scale (a) Saturn (R = 58 296 km) with its rings (from r = 70 000 km to r = 140 000 km), and its trabants Janus (r = 160 000 km), Mimas (r = 186 000 km), Enceladus (r = 238 000 km), Tethys (r = 295 000 km), Dione (r = 377 000 km), Rhea (r = 527 000 km), Titan (r = 1 222 000 km), and Hyperion (r = 1 481 000 km). Assume Saturn to be spherical, the rings and the trabant orbits circular and coplanar. (b) In a separate sketch and using a different scale, show Saturn, the orbit of Titan, and the orbits of Iapetus (r = 3 560 000 km) and Phoebe (r = 12 930 000 km). (c) Calculate the Roche limit, from

$$r_R = 2.4554 \left[\frac{\rho_M}{\rho_s} \right]^{1/3} R$$

where

ρ_M $= 680 \text{ kg/m}^3$ = mass density of Saturn

ρ_s $= 680 \text{ kg/m}^3$ = mass density of trabants

R $= 58\ 296$ km = radius of Saturn (sphere)

Appendix A

Space Data

Solar Time and Sidereal Time (365.2422 d = 366.2422 d sidereal time = 1 year)
24 h (solar time) = 24 h = 1440 min = 86 400 s = 24 h   3 min 56.555 s sidereal time
24 h sidereal time = 23 h 56 min 4.09 s (solar time) = 86 164.09 s (solar time)

Julian Date, Modified Julian Date
13 April 1982, 12 h UT: JD = 2 445 073.00 d, MJD = 45 072.50 d
13 April 1982, 24 h UT: JD = 2 445 973.50 d, MJD = 45 073.00 d
14 April 1982, 0 h UT: JD = 2 445 073.50 d, MJD = 45 073.00 d

Gravitational Constant (H. Cavendish, 1798)
$G = 6.670 \cdot 10^{-11}$ m^3/kg s^2

Stefan-Boltzmann Constant
$\rho - 5.67 \cdot 10^{-8}$ W/m^2 K^4

Absolute Temperature Zero
at -273.15°C (-459.67°F)

Solar Constant (in vicinity of Earth)
$S = 1353$ W/m^2

Speed of Light (in space)
$c = 299\ 792\ 458$ m/s (K.M. Evenson, 1980)

Astronomical Distances
1 light year = $9.46 \cdot 10^{15}$ m
1 parsec = 1 pc = 3.26 light years = $3.08 \cdot 10^{16}$ m

Speed of Sound (in dry air, standard pressure and temperature)
$c_s = 331.45$ m/s = 1193.22 km/h

Canonical Units
1 AU = 1 Astronomical Unit = Mean Distance Sun - Earth ($\approx$ 149 597 893 km)
1 MU = 1 Mass Unit = Mass of Sun ($\approx 1.989 \cdot 10^{30}$ kg)
1 DU = 1 Distance Unit = Radius of Attractor ($\approx$ 6371 km for Earth as Attractor)
1 TU = 1 Time Unit = Orbital Period for Orbit of 1 DU about Attractor ($\approx$ 84.347 min for Earth)

Astronomical Symbols

☉		Sun
☿		Mercury
♀		Venus
⊕		Earth
☽		Moon
♂		Mars
♃		Jupiter
♄		Saturn
♅		Uranus
♆		Neptune
♇		Pluto
☊	☊	Ascending Node
☋	☋	Descending Node
♈		First Point of Aries (= Vernal Equinox)

The Sun

Mass	$M = 1.989 \cdot 10^{30}$ kg
	$M = 1.34 \cdot 10^{29}$ lb s^2/ft
Volume	$V = 1.41 \cdot 10^{26}$ m^3
Surface	$A = 6.094 \cdot 10^{18}$ m^2
Gravitational Parameter	$\mu = 1.32715 \cdot 10^{20}$ m^3/s^2
Gravitational Attraction Constant at surface	$g = 273.8$ m/s^2
Escape Velocity	$v_p = 618$ km/s (from Sun surface)
Radius	$R = 696\,350$ km
Mean Density	$\rho = 1409$ kg/m^3
Schuler Period	$\tau_s = 167$ min

Mercury

Mass	$M = 0.3346 \cdot 10^{24}$ kg
Gravitational Parameter	$\mu = 22\,321$ km^3/s^2
Radius	$R = 2420$ km
Spin Period	$\tau = 58.6$ d
Mean Density	$\rho = 5620$ kg/m^3

Distance between Sun centre and Mercury centre

smallest:	$r_{PE} = 46.00 \cdot 10^6$ km
mean:	$a = 57.91 \cdot 10^6$ km
greatest:	$r_{AP} = 69.82 \cdot 10^6$ km

Eccentricity of Orbit around Sun	$\varepsilon = 0.2056$

Sidereal Period about Sun	$\tau \approx 87.97$ d
Inclination of Orbit Plane to Ecliptic	$7.00°$

Venus

Mass	$M = 4.822 \cdot 10^{24}$ kg
Gravitational Parameter	$\mu = 324\ 860$ km^3/s^2
Radius	$R = 6114$ km
Spin Period	$\tau = 243$ d
Mean Density	$\rho = 5090$ kg/m^3
Volume	$V = 0.959 \cdot 10^{20}$ m^3

Distance between Sun centre and Venus centre

$$\text{smallest: } r_{\text{PE}} = 107.48 \cdot 10^6 \text{ km}$$
$$\text{mean: } a = 108.21 \cdot 10^6 \text{ km}$$
$$\text{greatest: } r_{\text{AP}} = 108.94 \cdot 10^6 \text{ km}$$

Eccentricity of Orbit around Sun	$\varepsilon = 0.0068$
Sidereal Period about Sun	$\tau = 224.7$ d
Inclination of Orbit Plane to Ecliptic	$3°24'$

The Earth

Mass	$M = 5.976 \cdot 10^{24}$ kg
	$M = 4.0939 \cdot 10^{23}$ lb s^2/ft
Volume	$V = 1.083 \cdot 10^{21}$ m^3
Surface	$A = 509.088\ 842 \cdot 10^{12}$ m^2
Gravitational Parameter	$\mu = 398\ 601.19$ km^3/s^2
	$\mu = 1.407647 \cdot 10^{16}$ ft^3/s^2

Gravitation Attraction Constant on surface

standard value:	$g = 9.80665$ m/s^2 (incl. centrifugal component)
μ/R^2	$g = 9.82027$ m/s^2 (exl. centrifugal component)
at the equator:	$g = 9.81424$ m/s^2 (exl. centrifugal component)
	$g = 9.78032$ m/s^2 (incl. centrifugal component)
at the poles:	$g = 9.83218$ m/s^2

Near the Earth, g is reduced by approximately 1% for each 32 km height above the Earth's surface.

Radius	standard value:	$R = 6371$ km (Earth sphere)
	equatorial:	$R_e = a = b = 6\ 378.160$ km (Earth spheroid)
	polar:	$R_P = c = 6\ 356.775$ km

Flattening $\qquad f = \dfrac{1}{298.25} \qquad \varepsilon = 0.081\ 8$

Distance between Sun centre and Earth centre

$$\text{smallest: } r_{\text{PE}} = 147.1 \cdot 10^6 \text{ km (January-02)}$$
$$\text{mean: } a = 149.6 \cdot 10^6 \text{ km}$$
$$\text{greatest: } r_{\text{AP}} = 152.1 \cdot 10^6 \text{ km (July-03)}$$

Spin
$$\omega = \frac{2\pi}{23 \text{ h } 56 \text{ min } 4.09 \text{ s}} = \frac{2\pi}{86\ 164.09} = 72.921\ 159 \cdot 10^{-6} \text{ rad/s}$$
$$\tau = 23 \text{ h } 56 \text{ min } 4.09 \text{ s} = 23.934\ 469 \text{ h} = 96\ 164.09 \text{ s}$$

Tangential Surface Velocity at Equator	$v = 1674.369$ km/h $= 0.465\ 102$ km/s
Mean Density	$\rho = 5\ 514$ kg/m^3

Inclination of Equatorial Plane to Ecliptic

$$23° \, 27' = 23.45°$$

Eccentricity of Orbit around Sun $\quad$ $\varepsilon = 0.0167$

Orbital Period about Sun $\quad$ $\tau = 365.2422 \, d = 1 \, year = 1 \, a$

Orbital Velocity about Sun $\quad$ $v \approx 29.78 \, km/s$

Distance of a Synchronous Satellite $\quad$ $r = 42 \, 164 \, km$

Albedo $\quad$ 0.39

Coefficients of Zonal Spherical Harmonics

$$J_2 = 1082.682 \cdot 10^{-6}$$
$$J_3 = -2.538 \cdot 10^{-6}$$
$$J_4 = -1.593 \cdot 10^{-6}$$

Escape Velocity of Satellite $\quad$ $v_p = 11.2 \, km/s$ from Earth surface

Solar Wind Pressure $\quad$ $p = 9.02 \, \mu Pa$

(in vicinity of Earth, at equinox, for 100% reflecting surface, normal radiation)

Circular Velocity of Satellites $\quad$ $v_c = 7.91 \, km/s$ at Earth surface

Schuler Period $\quad$ $\tau_s = 84.347 \, min$

Inertia Moments $\quad$ $A = 80.152 \cdot 10^{30} \, km^2 kg$
(z-axis = North-South axis) $\quad$ $B = A$
$\quad$ $C = 80.416 \cdot 10^{30} \, km^2 kg$

Geosynchronous Satellite of Earth
in equatorial orbital plane;

Distance from Earth centre $\quad$ $r = 42 \, 164 \, km$

Angular Velocity about Earth $\quad$ $\omega = 72.92 \, 159 \cdot 10^{-6} \, rad/s$

Orbital Velocity $\quad$ $v = 3.075 \, km/s$

Orbital Period $\quad$ $\tau = 23 \, h \, 56.07 \, min$
$\quad = 23 \, h \, 56 \, min \, 4.09 \, s$
$\quad = 1436.07 \, min$
$\quad = 86 \, 164.09 \, s$

Heliosynchronous Satellite of Earth
in equatorial, ecliptic, or any other orbital plane

Distance from Earth centre $\quad$ $r = 498 \, 221 \, km$

Angular Velocity about Earth $\quad$ $\omega = 1.991 \cdot 10^{-6} \, rad/s$

Orbital Velocity $\quad$ $v = 0.99 \, 2 \, km/s$

Orbital Period $\quad$ $\tau = 876.173 \, h$
$\quad = 52 \, 570 \, min$
$\quad = 3 \, 154 \, 223.7 \, s$

The Moon

Mass $\quad$ $M = 73.50 \cdot 10^{21} \, kg$
$\quad M = 5.0 \cdot 10^{21} \, lb \, s^2/ft$

Gravitational Parameter $\quad$ $\mu = 4 \, 902.73 \, km^3/s^2$
$\quad \mu = 1.7313 \cdot 10^{14} \, ft^3/s^2$

Gravitational Attraction Constant at surface

$$g = 1.623 \, m/s^2$$
$$g = 5.31 \, ft/s^2$$

Radius $\quad$ $R = 1 \, 738 \, km$
$\quad R = 5.702 \cdot 10^6 \, ft$

Mean Density	$\rho = 3\ 344\ \text{kg/m}^3$

Inclination of Moon Orbital Plane to Ecliptic

$$5°\ 9'$$

Eccentricity of Moon Orbit about Earth	$\varepsilon = 0.0549$
Orbital Period about Earth	27 d 7 h 43 min = 27.321528 d
Orbital Velocity about Earth	$v \approx 1.022\ \text{km/s}$
	$\omega \approx 2.66 \cdot 10^{-6}\ \text{rad/s}$
Schuler Period	$\tau_s = 108.364\ \text{min}$
Spin Period	27 d 7 h 43 min = 27.321527 d

Moon Sol (from Sun culmination to Sun culmination)

$$29\ \text{d}\ 12\ \text{h}\ 44\ \text{min} = 29.5306\ \text{d}$$

Escape Velocity from lunar surface	$v_p = 2.375\ \text{km/s} = 7750\ \text{ft/s}$
Schuler Period	108.212 min

Circular Velocity of satellite at lunar surface

$$v_c = 1.679\ \text{km/s}$$

Surface Temperature	from +130°C to -150°C
Albedo	0.07

Distance of Earth centre to Moon centre

	smallest:	$r_{PE} = 363\ 300\ \text{km}$
	mean:	$a = 384\ 400\ \text{km}$
	greatest:	$r_{AP} = 405\ 500\ \text{km}$
Moon Orbit about Earth		$a = 384\ 400\ \text{km}\quad \varepsilon = 0.0549$
		$b = 383\ 365\ \text{km}$

Inclination of Lunar Orbital Plane to Earth Equatorial Plane
(varies and returns to the original value after 18.6 years)

$$i = \text{from } 18°\ 18'\ (1961, 1979, 1998) \text{ to } 28°\ 36'\ (1970, 1989, 2007)$$

Mass Ratio Moon/Earth	$\dfrac{M_{Moon}}{M_{Earth}} = \dfrac{1}{81.3} = 0.0123$
Inertia Moments	$A = 88.027 \cdot 10^{27}\ \text{km}^2\text{kg}$
	$B = 88.045 \cdot 10^{27}\ \text{km}^2\text{kg}$
	$C = 88.082 \cdot 10^{27}\ \text{km}^2\text{kg}$

(x-axis pointing to Earth, z-axis perpendicular to orbital plane)

Mars

Mass		$M = 6.42 \cdot 10^{23}\ \text{kg}$
Volume		$V = 1.6282 \cdot 10^{11}\ \text{km}^3$
Surface		$A = 1.4418 \cdot 10^8\ \text{km}^2$
Gravitational Parameter		$\mu = 42\ 828.3\ \text{km}^3/\text{s}^2$

Gravitational Attraction Constant at surface

	μ/R^2	$g = 3.733\ \text{m/s}^2$ (excluding centrifugal component)
	at the equator	$g = 3.710\ \text{m/s}^2$ (excluding centrifugal component)
		$g = 3.693\ \text{m/s}^2$ (including centrifugal component)
	at the poles	$g = 3.575\ \text{m/s}^2$
Radius	standard value:	$R = 3383\ \text{km}$ (Mars sphere)
	equatorial:	$R_e = 3393.5\ \text{km}$ (Mars spheriod)
	polar:	$R_P = 3362.1\ \text{km}$
Schuler Period		$\tau_s = 99.6\ \text{min}$

Spin
(Spin sense is the same $\omega = 0.255\ 175$ rad/h
as that of the Earth) $\tau = 24$ h 37 min 23 s $= 24.623$ h

Sol 24 h 37 min 22 s $= 24.659$ h

Tangential Surface Velocity at Equator $v = 864.743$ km/h $= 0.240206$ km/s

Mean Density $\rho = 3850$ kg/m^3

Inclination of Equatorial Plane to Orbital Plane
                                           $24.936°$

Inclination of Orbital Plane to Ecliptic

                                           $1.85°$

Inclination of Equatorial Plane to Ecliptic

                                           $23.086°$

Eccentricity of Orbit about Sun $\varepsilon = 0.093\ 4$

Orbital Period about Sun $\tau = 687$ d $= 669$ martian sols $= 1.881$ a

Orbital Velocity $v = 24.148$ km/s

Distance of Synchronous Satellite $r = 20\ 448$ km

Albedo 0.15

Coefficient of Zonal Spherical Harmonic

                                           $J_2 = -5.5 \cdot 10^{-6}$

Escape Velocity $v_p = 5.029$ km/s from Mars sphere surface

Circular Velocity of Satellite $v_c = 3.556$ km/s about Mars sphere surface

Distance between Sun centre and Mars centre
                        smallest: $r_{PE} = 206.6 \cdot 10^6$ km
                           mean: $a = 227.9 \cdot 10^6$ km
                        greatest: $r_{AP} = 249.2 \cdot 10^6$ km

Travel Time Earth-Mars   about 5.5 months

Trabants 2   Phobos   7.7 h period
                                           Deimos   30.3 h period

Mean Temperature at surface   -23°C

Atmosphere (in percent volume)   95% Carbon Dioxide
                                            3% Nitrogen
                                            1% Argon-40
                                           0.3% Oxygen
                                           plus traces of Argon 36, Krypton, and Xenon

Soil (in percent mass)   Al 2-7%   Cu < 0.5%
                                           Si 15-30%   Zn < 0.1%
                                           P < 10%   Ga < 0.03%
                                           S 0-6%   As < 0.02%
                                           Cl 0-3%   Se < 0.015%
                                           K 0-2%   Br < 0.015%
                                           Ca 3-8%   Rb < 0.01%
                                           Ti 0.5-2%   Sr < 0.02%
                                           V < 3%   Y < 0.02%
                                           Cr < 5%   Zr < 0.02%
                                           Mn < 7%   Nb < 0.025%
                                           Fe 14±2%   Mo < 0.05%
                                           Co < 7%   Tc to U < 2%
                                           Ni < 5%

Jupiter

Mass	$M = 1.900 \cdot 10^{27}$ kg
Volume	$V = 1.456 \cdot 10^{23}$ m^3
Gravitational Parameter	$\mu = 126.7 \cdot 10^6$ km^3/s^2
Gravitational Attraction Constant	$g = 25.238$ m/s^2 (without centrifugal component)
Radius	$R_e = 71\,818$ km

Flattening

$$f = \frac{1}{15.2}$$

Spin

$$\omega = 0.633 \text{ rad/h}$$
$$\tau = 9\text{ h}\ \ 55.5\text{ min} = 9.925\text{ h}$$

Tangential Surface Velocity at Equator

$$v = 12.5 \text{ km/s}$$

Schuler Period $\tau_s = 179$ min

Coefficients of Zonal Spherical Harmonics

$$J_2 = 1.475 \cdot 10^{-2}$$
$$J_4 = -0.058 \cdot 10^{-2}$$
$$J_6 = 0.005 \cdot 10^{-2}$$

Mean Density $\rho = 1310$ kg/m^3

Distance between Sun centre and Jupiter centre

$$\text{smallest:}\ r_{PE} = 740.6 \cdot 10^6 \text{ km}$$
$$\text{mean:}\ a = 778.3 \cdot 10^6 \text{ km}$$
$$\text{greatest:}\ r_{AP} = 816 \cdot 10^6 \text{ km}$$

Eccentricity of Orbit around Sun	$\varepsilon = 0.0485$
Orbital Period about Sun	$\tau = 11.86$ a
Travel Time, Earth to Jupiter:	about 2.7 years

Saturn

Mass	$M = 568.3 \cdot 10^{24}$ kg
Volume	$V = 83.445 \cdot 10^{21}$ m^3
Surface	
Gravitational Parameter	$\mu = 37.9 \cdot 10^6$ km^3/s^2

Gravitational Attraction Constant on surface

$$g = 9.12 \text{ m/s}^2$$

Radius

$$\text{standard value:}\ R = 58\,296 \text{ km} \qquad \text{(Saturn sphere)}$$
$$\text{equatorial:}\ R_e = a = b = 60\,335 \text{ km} \qquad \text{(Saturn spheroid)}$$
$$\text{polar:}\quad c = 54\,422 \text{ km}$$

Flattening	$f = 0.098$
Spin	$\omega = 0.614$ rad/h
	$\tau = 10\text{ h } 14\text{ min} = 10.23\text{ h}$
Tangential Surface Speed at Equator	$v = 37\,045$ km/h $= 10.29$ km/s
Mean Density	$\rho = 680$ kg/m^3 (lighter than water!)

Inclination of Orbital Plane to Elliptic

$$2°29' = 29.48°$$

Inclination of Saturn Axis to Orbital Normal

$$26°44' = 26.73°$$

Distance between Sun centre and Saturn centre

$$\text{smallest:}\ r_{PE} = 1347.6 \cdot 10^6 \text{ km}$$
$$\text{mean:}\ a = 1427.0 \cdot 10^6 \text{ km}$$

$$\text{greatest:}\quad r_{AP} = 1506.4 \cdot 10^6 \text{ km}$$

Eccentricity of Orbit	$\varepsilon = 0.0556$
Orbital Period about Sun (Sidereal Period)	$\tau = 29.46$ a
Mean Orbital Velocity	$v = 9.6$ km/s
Distance of Synchronous Satellite	$r = 50\ 698$ km
Albedo	0.41
Escape Velocity	$v_p = 33.1$ km/s

Uranus (1781, W. Herschel)

(Voyager 2, $\uparrow$ 1977-September; Jupiter rendezvous 1979-July-9; Saturn 1981-August-25; Uranus 1986-January-31; Neptune 1989-August)

15 trabants (5 outer trabants: Miranda (innermost, 500 km diameter), Ariel (1160 km diameter), Umbriel (1190 km diameter), Titania (1600 km diameter), Oberon (1600 km diameter) on orbit of 583 400 km radius, plus 10 small ones within the rings.

11 rings

Mass	$M = 86.65 \cdot 10^{24}$ kg
Volume	$V = 54.8 \cdot 10^{21}$ m^3
Surface	$A = 8.1 \cdot 10^{15}$ m^3
Gravitational Parameter	$\mu = 5\ 780\ 000$ km^3/s^2

Gravitational Attraction Constant (at equator)

$$g = 9.7 \text{ m/s}^2$$

Radius (at equator)	$R_e = 23500$ km
Flattening	$f = \dfrac{1}{18} = 0.056$
Spin	$\omega = 0.3645$ rad/h
	$\tau = 17.24$ h

Rotational Axis located in its orbital plane

Mean Density	$\rho = 1580$ kg/m^3

Distance between Sun centre and Uranus centre

smallest:	$r_{PE} = 2734 \cdot 10^6$ km
mean:	$a = 2870 \cdot 10^6$ km
greatest:	$r_{AP} = 3006 \cdot 10^6$ km
Eccentricity of Orbit around Sun	$\varepsilon = 0.0472$
Orbital Period about Sun	$\tau = 84.015$ years

Travel Time, Earth to Uranus about 16.03 years (only 9 years for Voyager 2, due to energy pick-up at Jupiter and Satur

Magnetic field (1 gauss on surface of Uranus)

Atmosphere of Hydrogen, Ammonia, and Methane

Neptune (1846, J.G. Galle)

2 trabants (Triton (3800 km diameter, larger than the Moon!), Nereide), rings?

Mass	$M = 102 \cdot 10^{24}$ kg
Volume	$V = 69 \cdot 10^{21}$ m^3
Surface	$A = 6.21 \cdot 10^{15}$ m^2
Gravitational Parameter	$\mu = 6\ 800\ 000$ km^3/s^2
Gravitational Attraction Constant	$g = 10.8$ m/s^2
Radius	$R = 25\ 000$ km

Spin	$\omega = 0.3977$ rad/s $\tau = 15.8$ h
Mean Density	$\rho = 2000$ kg/m^3

Distance between Sun centre and Neptune centre

smallest:	$r_{PE} = 4458 \cdot 10^6$ km
mean:	$a = 4497 \cdot 10^6$ km
greatest:	$r_{AP} = 4535 \cdot 10^6$ km
Eccentricity of Orbit around Sun	$\varepsilon = 0.0086$
Orbital Period about Sun	$\tau = 167.79$ years
Travel Time, Earth to Neptune	about 30.56 years (only some 11 years for Voyager 2, due to energy pick-up at Jupiter, Saturn, and Uranus)

Pluto

(1930, Clyde Tombaugh)

Mass	Pluto and Charon $M = 15 \cdot 10^{21}$ kg Pluto alone 87% $13.05 \cdot 10^{21}$ kg Charon alone 13% $1.95 \cdot 10^{21}$ kg
Radius	$R = 1100$ km (1986, Manfred Pakull, Klaus Reinsch)
Spin Period	$\tau = 6.4$ d

Distance between Sun centre and Pluto centre

smallest:	$r_{PE} = 4439 \cdot 10^6$ km
mean:	$a = 5946 \cdot 10^6$ km
greatest:	$r_{AP} = 7453 \cdot 10^6$ km
Eccentricity of Orbit around Sun	$\varepsilon = 0.253$
Orbital period about Sun	$\tau = 247.7$ a
Mean Orbital Speed	$v = 4.7$ km/s
Trabant	Charon (1978, James Christy) Radius $R = 580$ km, Mass $M = 1.95 \cdot 10^{21}$ kg, in a circular orbit of 19 400 km radius about Pluto

Appendix B

Engineering Data

Systems of Units

System	Basic Units				Standard Gravitational Attraction Constant
	Length	Force	Time	Mass	
MKSA (SI)	Metre, m		Second, s	Kilogram, kg	$g = 9.80665 \text{ m/s}^2$
FPS (U.S.A.)	Foot, ft	Pound, lb	Second, s		$g \approx 32.2 \text{ ft/s}^2$

$$\text{Force in MKSA system:} \quad \text{Newton, N} = \frac{\text{m kg}}{\text{s}^2}$$

$$\text{Mass in FPS system:} \quad \text{Slug, slug} = \frac{\text{lb s}^2}{\text{ft}}$$

Conversion Factors:

$$1 = 0.3048 \frac{\text{m}}{\text{ft}} \qquad 1 \approx 4.448 \frac{\text{N}}{\text{lb}} \qquad 1 \approx 14.593 \frac{\text{kg}}{\text{slug}}$$

Larger and Smaller Units in the SI (Système International)

	Factor	Prefix	Symbol
	10^{18}	Exa-	E
	10^{15}	Peta-	P or PA
	10^{12}	Tera-	T
Larger	10^{9}	Giga-	G
Units	10^{6}	Mega-	M
	10^{3}	Kilo-	k or K
	10^{2}	Hecto-	h or H
	10^{1}	Deka-	da or D or dk
	10^{-1}	deci-	d
	10^{-2}	centi-	c
	10^{-3}	milli-	m
Smaller	10^{-6}	micro-	μ
Units	10^{-9}	nano-	n
	10^{-12}	pico-	p
	10^{-15}	femto-	f
	10^{-18}	atto-	a

Example:

$$r_{\text{Sun–Earth}} = 150 \text{ Gm} = 150\,000\,000 \text{ km} = 150 \cdot 10^9 \text{ m}$$

Material Properties

Material	Density ρ kg/m^3	Young's Modulus E MPa	Speed of Sound* c m/s	Tensile Strength σ_{ult} MPa
Air (0° C)	1.293	0.14	331	
Aluminum (2014-T6)	2 690	55 000	5 240	240
Beryllium Copper (C17200)	8 290	120 000	4 410	1 500
Brass (naval)	8 450	100 000		420
Brick	1 800	17 300	3 600	240
Bronze (phosphor ASTM B 159)	8 930	100 000		700
Concrete	2 400	18 300	3 200	175**
Copper	9 810	84 900	3 580	210-450
Dacron (polyester)	1 380	13 800		1 120
Glass	2 590	60 000	5 560	90(900**)
Gold	19 300	59 100	2 030	140
Ice	900			13**
Iron (cast, gray, No. 60)	7 210	130 000	5 170	420
Kapton (polyimide film)	1 400	3 000	4.8	172
Kevlar 29 (aramid fibre)	1 440	12 000		2 760
Lead	11 370	13 200	1 250	14
Magnesium (AZ80A-T5)	1 800	43 000	1 310	380
Mercury	13 570		100	
Nylon (728)	1 140	5 520		985
Oil	90	830		
Rubber	920	2	54	15
Soil, wet	1 760			
dry	1 280			
Steel (Spring alloy SAE 4068)	7 830	200 000	5 850	1 720
Titanium	3 080	110 000	6 420	800
Water, fresh	1 000	2 000	1 400	
salt	1 030	2 000	1 560	
Wood, oak	800	12 000	3 380	490**
pine	480	14 000	4 180	590**

$$* \; c = \sqrt{\frac{1-v}{(1+v)(1-2v)}\frac{E}{\rho}} \qquad ** \text{ in compression}$$

Series

$$\frac{1}{(1+x)^2} = 1 - 2x + 3x^2 - 4x^3 + 5x^4 - \ldots$$

$$\frac{1}{(1-x)^2} = 1 + 2x + 3x^2 + 4x^3 + 5x^4 + \ldots$$

$$\sin x = x - \frac{1}{3!}x^3 + \frac{1}{5!}x^5 - \frac{1}{7!}x^7 + \ldots$$

$$\cos x = 1 - \frac{1}{2!}x^2 + \frac{1}{4!}x^4 - \frac{1}{6!}x^6 + \ldots$$

$$\tan x = x + \frac{1}{3}x^3 + \frac{2}{15}x^5 + \frac{17}{315}x^7 + \ldots$$

$$e^x = 1 + x + \frac{1}{2!}x^2 + \frac{1}{3!}x^3 + \frac{1}{4!}x^4 + \ldots$$

$$\ln(1+x) = x - \frac{1}{2}x^2 + \frac{1}{3}x^3 - \frac{1}{4}x^4 + \frac{1}{5}x^5 - \ldots$$

$$(1 \pm x)^n = 1 \pm nx + \frac{n(n-1)}{2!}x^2 \pm \frac{n(n-1)(n-2)}{3!}x^3 + \ldots$$

Expansions of $\dfrac{R}{d} = \sqrt{1 - 2\dfrac{r}{d}\cos\alpha + \dfrac{r^2}{d^2}}$

$$\frac{R}{d} = 1 - \frac{r}{d}\cos\alpha + \frac{1}{2}\frac{r^2}{d^2}(1 - \cos^2\alpha) + \ldots$$

$$\frac{R^2}{d^2} = 1 - 2\frac{r}{d}\cos\alpha + \frac{r^2}{d^2}$$

$$\frac{R^3}{d^3} = 1 - 3\frac{r}{d}\cos\alpha + \frac{3}{2}\frac{r^2}{d^2}(1 + \cos^2\alpha) + \ldots$$

$$\frac{R^4}{d^4} = 1 - 4\frac{r}{d}\cos\alpha + 2\frac{r^2}{d^2}(1 + 2\cos^2 d) + \ldots$$

$$\frac{d}{R} = 1 + \frac{r}{d}\cos\alpha + \frac{1}{2}\frac{r^2}{d^2}(3\cos^2\alpha - 1) + \ldots$$

$$\frac{d^2}{R^2} = 1 + 2\frac{r}{d}\cos\alpha + \frac{r^2}{d^2}(4\cos^2\alpha - 1) + \ldots$$

$$\frac{d^3}{R^3} = 1 + 3\frac{r}{d}\cos\alpha + \frac{3}{2}\frac{d^2}{r^2}(5\cos^2\alpha - 1) + \ldots$$

$$\frac{d^4}{R^4} = 1 + 4\frac{r}{d}\cos\alpha + 2\frac{r^2}{d^2}(6\cos^2\alpha - 1) + \ldots$$

Appendix C

Nomenclature

Symbols

A	=	area
		principal inertia moment about x-axis
AP	=	apoapsis
B	=	burnout angle
B	=	principal inertia moment about y-axis
C	=	mass centre
C	=	centrifugal force
C	=	principal inertia moment about z-axis
D	=	special parameter for parabola
E	=	total energy
E	=	eccentric anomaly
E$'$	=	eccentric anomaly for hyperbola
F	=	force
F	=	focus
F_o	=	occupied focus
F_v	=	vacant focus
G	=	universal gravitational constant
H	=	angular momentum
I_C	=	inertia moment about O-C axis
J	=	coefficient of the zonal spherical
K	=	Kepler force
L	=	libration point
M	=	moment
M	=	mean anomaly
M$'$	=	mean anomaly for hyperbola
M	=	mass
N	=	North
O	=	origin
P	=	point
P_n	=	Legendre polynomial of nth degree
PE	=	periapsis
R	=	reference point
R	=	radius
		nominal radius of master mass sphere
R'	=	radius in case of hyperbola
S	=	impulse
S	=	satellite
		star
		South
T	=	kinetic energy
		temperature

T	=	transfer point
U	=	potential energy
V	=	volume
X, Y, Z	=	coordinates
$\mathbf{a}$	=	acceleration
a	=	semi-major axis of ellipse distance side of spherical triangle, line angle opposite α radius
a'	=	semi-axis of hyperbola
$\mathbf{a}$	=	year
b	=	side of spherical triangle, line angle opposite β semi-minor axis of ellipse distance radius
b'	=	second semi-axis of hyperbola
b'_c	=	collision radius
c	=	side of spherical triangle, line angle opposite γ
c_d	=	drag coefficient
$\mathbf{d}$	=	distance vector
d_E	=	distance of Earth
d_M	=	distance of Moon
$\mathbf{e}$	=	specific mechanical energy
e	=	linear eccentricity of ellipse
e'	=	linear eccentricity of hyperbola
$\mathbf{e}_x$	=	unit vector in x-direction
$\mathbf{f}$	=	perturbation force
f	=	flattening
g	=	gravitational attraction constant on surface of master mass
g_μ	=	micrograviational attraction coefficient
$\mathbf{h}$	=	specific angular momentum
i	=	inclination angle of orbital plane
j	=	$\sqrt{-1}$
m	=	mass
n	=	mean angular speed
p	=	semi-parameter, semi-latus rectum solar pressure
q	=	perifocal distance
$\mathbf{r}$	=	position vector
r	=	radial distance polar coordinate
r_o	=	initial distance
r_{IS}	=	radius of influence sphere
r_R	=	Roche limit
s	=	arc length
t	=	hour angle time temperature in °C

t_{PE}	=	epoch of satellite passage through periapsis
t_λ	=	local time
u	=	orbit angle (from ascending node)
v	=	velocity
v_c	=	speed on circular orbit
v_p	=	escape speed
v_∞	=	hyperbolic excess speed
x,y,z	=	coordinates
x	=	universal variable
α	=	direction angle with respect to x-axis plane angle in spherical triangle, opposite side a angle between tangent and radial right ascension first Lambert parameter
β	=	direction angle with respect to y-axis plane angle in spherical triangle, opposite side b geocentric latitude flight path angle, angle between tangent and transverse second Lambert parameter
β_o	=	initial flight path angle
γ	=	direction angle with respect to z-axis plane angle in spherical triangle, opposite side c
δ	=	declination deflection angle
ε	=	numerical eccentricity
θ	=	true anomaly polar angle latitude
θ_∞	=	hyperbolic asymptote angle
κ	=	tilt angle
λ	=	geocentric longitude
μ	=	gravitational parameter
ξ,η,ζ	=	coordinates
ρ	=	density radial distance radius of curvature
τ	=	Kepler period
τ_{PE}	=	apsidal period
τ_s	=	Schuler period
τ_Ω	=	nodal period
$\tau_{2\pi}$	=	absolute period
ϕ	=	angle between two vectors polar angle latitude angle
Ω	=	angular velocity of coordinate system orbital angular velocity on circular orbits right ascension of ascending node
ω	=	orbit angle of periapsis angular velocity

Subscripts

AP	=	apoapsis
C	=	centre, mass centre
E	=	Earth
IS	=	influence sphere
M	=	Moon
P	=	Pole
PE	=	periapsis
R	=	reference point
T	=	transfer
S	=	satellite
		Sun
c	=	circular
e	=	equator
en	=	entry
ex	=	exit
l	=	launch
o	=	original, initial
p	=	parabolic, escape
pert	=	perturbation
r	=	radius
n	=	normal
t	=	tangential
syn	=	synodic
θ	=	transverse

Superscripts

S″	=	supersatellite point (projection onto celestial sphere)
S′	=	subsatellite point (projection onto master sphere)

Appendix D

Answers to Selected Problems

Chapter 1

1-1	(a) $h = 56\ 115\ \text{km}^2/\text{s}$ (b) $h = 390\ 846\ \text{km}^2/\text{s}$

1-2	ε	b
	0	5
	0.2	4.899
	0.4	4.583
	0.6	4
	0.8	3
	1	0

1-3	Sun	$\dfrac{F_{SUN}}{F_{EARTH}} = 2.185$

1-8
$$\ddot{r} + r\dot{\theta} = \frac{\mu}{r^2}$$
$$r\ddot{\theta} + 2\dot{r}\dot{\theta} = 0$$

1-9
$$[\omega] = \begin{bmatrix} 0 & -\omega_z & \omega_y \\ \omega_z & 0 & -\omega_x \\ -\omega_y & \omega_x & 0 \end{bmatrix}$$

Chapter 2

2-1	$\tau_1 = 97.141$ min	$\tau_2 = 97.877$ min	$\tau_3 = 98.617$ min
2-2	Sun = 167 min Mars = 99.74 min	Earth = 84.35 min Jupiter = 177 min	Moon = 108.1 min
2-3	(a) Sun = 436 km/s Earth = 7.91 km/s Moon = 1.681 km/s Mars = 3.556 km/s Jupiter = 42.00 km/s	(b) Sun = 616.6 km/s Earth = 11.19 km/s Moon = 2.38 km/s Mars = 5.03 km/s Jupiter = 59.4 km/s	
2-4	$v = 6.63$ km/s		

2-5

| True Anomaly | | Eccentric Anomaly | | Mean Anomaly | |
| θ | | E | | M | |
°	rad	°	rad	°	rad
0	0	0	0	0	0
45°	$\pi/4$	23.402°	0.408	9.723°	0.167
90°	$\pi/2$	53.130°	0.927	25,611°	0.447
126.87°		90°	$\pi/2$	55.622°	0.971
135°		100.721°	1.758	66.948°	1.168
147.69°		119.828°	2.091	90°	1.570
150°		123.626°	2.157	94.962°	1.657
180°	π	180°	π	180°	π
270°		306.876°	5.356	334.378°	5.836
360°	2π	360°	2π	360°	2π

2-6

$$\omega = \sqrt{\frac{\mu}{R^3}}$$

2-7

(a) $\ddot{\theta} + \frac{g}{l}\theta = \frac{1}{l}(l-R)\ddot{\alpha}$

(b) $l = R$

(c) $\tau = 84.4$ min

2-8

(a) $W = 977.725$ N

(b) $W = 982.068$ N

(c) $W = 979.868$ N

2-9

(a) $E = -62\,565$ kJ

(b) $E = -30\,197$ kJ

(c) $E = -4\,727$ kJ

2-10

$t = 0.5$ s　roughly

2-11

$\delta = 685$ m

2-13

(a) $\tau = 213.71$ h

(b) $\tau = 302$ h

2-14

Orbit	r_{PE}	p	b	a
A	6700	6700	6700	6700
B	6700	7370	7407	7444
C	6700	8040	8205	8375
D	6700	10 050	11 605	13 400

2-15

(a) 182 orbits　　(b) 0.988°

2-16

(a) $T = 3.152$ N　　(b) $d = 0.002$ mm　　(c) $m_c = 0.045$ kg

Chapter 3

3-1

$$\varepsilon = \left| \left[1 - \frac{r_o}{a} \right] \sqrt{1 + \left[2 - \frac{r_o}{a} \right] \frac{r_o}{a} \sin^2\beta_o \left[1 - \frac{r_o}{a} \right]^{-2}} \right|$$

3-2

β_o	ε_1					
	$v_o = 0$	$0.9\sqrt{\mu/r_o}$	$1\sqrt{}$	$1.1\sqrt{}$	$\sqrt{2}\sqrt{}$	$2\sqrt{\mu/r_o}$
0	-1	-0.19	0	0.21	1	3
10°	-1	-0.55	0.173	0.27	1	2.959
20°	-1	-0.386	0.342	0.395	1	2.839
30°	-1	-0.526	0.5	0.532	1	2.646
45°	-1	-0.720	0.707	0.722	1	2.236
90	-1	-1	1	1	1	1

3-3 $\varepsilon = |\sin\beta_o|$

3-4 $\varepsilon = |\sin\beta_o| = 0.173\ 648$ $a = 7000\ \text{km}$
$v_o = 7.546\ \text{km/s}$ $\theta_o = 100°$

3-6 $\beta_o = 10.29°$

3-7 $a = 21\ 372\ \text{km}$ $\varepsilon = 0.7373$ $\theta_o = 72.7°$
$2\varepsilon a = 31\ 515\ \text{km}$

3-8 (a) $a' = 1\ 139\ 674\ \text{km}$ (c) $\theta_o = 39.871°$
(b) $\varepsilon = 1.0062$ (d) $2\varepsilon a' = 2\ 293\ 480\ \text{km}$
(e) hyperbola

Chapter 4

4-1 $a_2 = 8155\ \text{km}$ $\varepsilon_2 = 0.169\ 754$ $b_2 = 8036\ \text{km}$
$\theta_2 = 84.092°$

4-2 $\kappa = 2.278°$ $\Delta\Omega = 1.986°$ $i_2 = 61.479°$ $u_2 = 49,029°$

4-3 $x = 0.302\ 648$

4-4 $x = 0.064\ 177\ 777$ $y_{\max} = 0.536\ 258\ 305$

4-5 (a) $a_2 = 19\ 054\ \text{km}$ (b) $a_2 = 15\ 421.26\ \text{km}$
      $\varepsilon_2 = 0.63$ $\varepsilon_2 = 0.36$
      $E_{new} - E_{old} = 755.2\ \text{kJ}$ $E_{new} - E_{old} = 246.4\ \text{kJ}$

4-6	$\Delta v = 0.027\ 599$ km/s $\\ F = 3.384 \cdot 10^{19}$ kN $\\ \Delta E = 1.984 \cdot 10^{21}$ MJ $\\ \tau_1 = 27.45$ d $\\ \tau_2 = 27.74$ d	a tremendous amount, impossible to supply		
4-7	$\kappa = 2.949°$ $\\ \Delta\Omega = 16.442°$ $\\ i_2 = 7.385°$ $\\ u_2 = 28.648°$			
4-8	(a) $\tau_T = 0.989$ h	(b) $\Delta V = 0.888$ km/s	(c) $\Delta E = 898$ MJ	
4-9	(a) $\Delta V = 0.414\ v_o$	(b) $\Delta V = -v_o$		

4-20

(i) (a) $\Delta v_1 = 1.391$ km/s (ii) $\Delta v_1 = 2.343$ km/s
(b) $\Delta v_2 = 1.478$ km/s $\Delta v_2 = 1.028$ km/s
(c) $\Delta v = 2.869$ km/s $\Delta v = 3.371$ km/s
(d) $\tau_T = 5.26$ h $\tau_T = 6.16$ h
(e) $\Delta T_1 = 13\,285$ MJ $\Delta T_1 = 14\,104$ MJ
(f) $\Delta T_2 = 3\,452$ MJ $\Delta T_2 = 4\,727$ MJ
(f) $\Delta T = 16\,736$ MJ $\Delta T = 16\,736$ MJ

4-21

(a) $\Delta v = 15.708$ km/s $\tau_T = 30.56$ a
(b) $\Delta v = 15.218$ km/s $\tau_T = 267.34$ a
(c) $\Delta v = 14.588$ km/s $\tau_T = \infty$

Chapter 5

5-1

$$P_5(x) = \frac{63}{8}\,x^5 - \frac{70}{8}\,x^3 + \frac{15}{8}\,x$$

5-2

$g^* = 9.814$ m/s² at equator, $g^* = 9.832$ m/s² at poles

5-3

$$J_2 = \frac{2C - B - A}{2MR_e^2} \qquad J_2 = \frac{C-A}{MR_e^2} \text{ when } B = A$$

$$C - A = 0.26 \cdot 10^{30} \text{ km}^2\text{kg}$$

5-4

(a) $$U = -\frac{\mu m}{r}\left\{1 + \frac{A(1-3\cos^2\alpha) + B(1-3\cos^2\beta) + C(1-3\cos^2\gamma)}{2Mr^2}\right\}$$

(b) $F_r = -10\,276\,432$ kN
$F_\alpha = -269\,686$ kN

5-5

$$U = -\frac{\mu m}{r}\left\{1 + \frac{1}{4}\,\frac{R^2}{r^2}\,(1 - 3\sin^2\delta)\right\}$$

$$r = -\frac{\mu m}{U}\left\{1 + \frac{1}{4}\,\frac{R^2}{r^2}\right\}$$

approx. $\qquad r = -\frac{\mu m}{U}\left\{1 + \frac{1}{4}\,\frac{R^2U^2}{\mu^2 m^2}\right\}$ for $\delta = 0°$

approx. $\qquad r = -\frac{\mu m}{U}\left\{1 - \frac{1}{2}\,\frac{R^2U^2}{\mu^2 m^2}\right\}$ for $\delta = 90°$

5-6

(a) $$U = -\frac{\mu m}{r}\left\{1 - \frac{1}{8}\,\frac{R^2}{r^2}\,(3\sin^2\delta - 1)\right\}$$

(b) $$r = -\frac{\mu m}{U}\left\{1 + \frac{1}{8}\,\frac{R^2U^2}{\mu^2 m^2}\right\}$$

$$r = -\frac{\mu m}{U}\left\{1 - \frac{1}{4}\,\frac{R^2U^2}{\mu^2 m^2}\right\}$$

5-7
$$U = -\frac{GMm}{r}\left[1 - \frac{1}{2}\frac{R^2}{r^2}\frac{31}{240}(3\sin^2\delta - 1)\right]$$

5-9
(a) $0.000\,006\,111 \cdot 10^{-10}\%$
(b) $163.6 \cdot 10^{15}$ years
(c) $0.987 \cdot 10^{15}$ km²kg

5-10
$$\Delta H = -\frac{2\pi C}{\tau}\Delta\tau + \tfrac{1}{2}m\sqrt{\frac{\mu}{r}}\,\Delta r = 0$$

Chapter 6

6-2
$i = 30.08°$

6-3
$\Delta\varepsilon = 0, \quad \Delta\Omega = 0, \quad \Delta\omega = 0, \quad \Delta i = 0$
$\Delta a = 6.798$ km

6-4
$\Delta\varepsilon = 0 \quad \Delta\Omega = 2.383° \quad \Delta\omega = -2.036°$
$\Delta i = 1.19° \quad \Delta a = 0$

6-5
$\Delta\varepsilon = 0 \quad \Delta\Omega = 3.325° \quad \Delta\omega = -2.889°$
$\Delta i = -0.257° \quad \Delta a = 0$

6-6
$\Delta\Omega = 1.459° \qquad \Delta\omega = -0.263°$
$\Delta i = 0.729° \qquad \Delta\varepsilon = \Delta\omega = 0$

6-7
$\Delta a = 122.33$ km $\qquad \Delta\omega = 1.16°$
$\Delta\varepsilon = 0.004\,873 \qquad \Delta i = -0.218°$
$\Delta\Omega = 0.069°$

6-8
(a) $\Delta a = \dfrac{p^2}{\mu}\dfrac{2p}{(1-\varepsilon^2)^2}\dfrac{2\pi}{\sqrt{1-\varepsilon^2}}\dfrac{f_\theta}{m}$
(b) $\Delta a = 23.993$ km

6-9
(a) $\dfrac{da}{d\theta} = \dfrac{p}{\mu}\dfrac{2p^2}{(1-\varepsilon^2)^2}\dfrac{\sqrt{1-2\varepsilon\cos\theta + \varepsilon^2}}{(1+\varepsilon\cos\theta)^2}\dfrac{f_t}{m}$

(b) $\Delta a = \dfrac{4\pi a^3}{\mu}\dfrac{f_t}{m}$

(c) $\Delta a = 25.22$ km

6-10
$\Delta a = 2.261$ km $\qquad \Delta\omega = 0.001\,636°$
$\Delta\varepsilon = 0.000\,207 \qquad \Delta i = -0.008\,076°$
$\Delta\Omega = 0.540°$

6-11
(a) $f_o = 50.625$ N $\qquad\qquad$ (b) $\varepsilon = 0.51738$
$f_{90} = 15$ N
$f_{180} = 1.875$ N
$f_{270} = 15$ N
$f_{360} = 50.625$ N

Chapter 7

7-1	(a) $\Delta a = \dfrac{4\pi a^2 b}{\mu}\,\dfrac{f_\theta}{m}$ (b) $\Delta a = 12.61$ km (c) $\dfrac{f_\theta}{k} = 0.0001$
7-2	$\dot{\Omega} = 1.403$ deg/day
7-3	$\Delta a = 0$
7-4	$\omega = -0.237°$
7-5	$\omega = 138.2°$
7-6	751 km

Chapter 8

8-1
(a) $\tau = 8145.592$ s $= 135.760$ min
(b) $\tau_{2\pi} = 8147.499$ s $= 135.792$ min
(c) $\tau_\Omega = 8147.499$ s $= 135.792$ min
(d) $\tau_{PE} = 8143.685$ s $= 135.728$ min

8-2
(a) $\tau = 121.701$ min
(b) $\tau_{2\pi} = 121.729$ min
(c) $\tau_\Omega = 121.721$ min
(d) $\tau_{PE} = 121.673$ min

8-3
(a) $\tau = 106.538\ 906$ min $= 6392.327\ 4353$ s
(b) $\tau_{2\pi} = 6393.438\ 086$ s
(c) $\tau_\Omega = 6392.327\ 435$ s
(d) $\tau_{PE} = 6391.216\ 784$ s

8-4
$$\tau_{2\pi} = \tau\left[1 \ - \ \frac{3}{8}\,J_2\,\frac{R_e^2}{a^2\sqrt{1-\varepsilon^2}}\,(3\cos^2 - 1)\right]$$
(a) $\tau_{2\pi} = 6562.2032 - 2.3724$ s
(b) $\tau_{2\pi} = 6262.2032 - 2.4084$ s

8-5
(a) $\tau = 12\ 846$ s $= 3.568$ h
(b) $\tau_{2\pi} = 12\ 902$ s
(c) $\tau_\Omega = 12\ 901$ s
(d) $\tau_{PE} = 12\ 790$ s

Chapter 9

9-1 $\Delta a = 0.997$ m

9-2
(a) $\Delta a = -32.236$ m
(b) $\Delta\varepsilon = -0.000\ 004$
(c) $\Delta r_{PE} = 0$
    $\Delta v_{AP} = 0.045$ m/s
(d) $\Delta r_{AP} = -64.472$ m
(e) $\Delta V_{PE} = -0.015$ m/s
(f) $\Delta V_{AP} = 0.045$ m/s
    $\Delta r_{AP} = -64.472$ m

9-3
(a) $\Delta a = -135.6$ m   when $\rho = 10^{-9}$ kg/m^3
(b) $\Delta a = -1.356$ m   when $\rho = 10^{-11}$ kg/m^3

9-4	accel $= 1.118\,(10^{-5})$ m/s^2

9-5

(a) $v = 7800$ m/s
(b) $v = 7729$ m/s
(c) $D = 21.676$ N
(d) $W = 473$ N
(e) $C = 460$ N

(f) $T = 25.219$ N
(g) $\beta = 30.739°$
(h) $l = 117.389$ km
(i) $m = 132.764$ kg
(j) $\sigma = 32.11$ MPa
(k) $F = 25.219$ N

9-6 $c_d = 2.01$

9-7

(a) $D = 2.4$ N
(b) $F = 0.136$ N
(c) $\delta = 34$ m

Chapter 10

10-1

$$r_{\mathrm{IS}} = \left[\frac{\mu_B}{\mu_A}\right] 0.4\,d \qquad \text{Earth: } r_{\mathrm{IS}} = 927\,133 \text{ km}$$

Mercury: $r_{\mathrm{IS}} = 110\,446$ km　　Jupiter: $r_{\mathrm{IS}} = 48\,684\,664$ km
Venus:　　$r_{\mathrm{IS}} = 618\,963$ km　　Saturn: $r_{\mathrm{IS}} = 54\,712\,268$ km

10-2

$$y^{-3} = 1 + 3ax - \frac{3}{2}x^2 + \frac{15}{2}a^2x^2$$
$$y^{-4} = 1 + 4ax - 2x^2 + 12\,a^2x^2$$

10-3

$v_\infty = 2.35$ km/s　　　　　　$\delta = 88.213°$

estimate $\dfrac{d}{v} = t = 136.4$ h upper bound

Lambert $t = 129$ h lower bound
eccentric anomaly $t = 129.666$ h

10-4　　ahead $\Delta T = -53\,200$ MJ　　　　behind $\Delta T = 182\,400$ MJ

10-5

(a) $a_1 = 464 \cdot 10^6$ km
　　$\varepsilon_1 = 0.676\,725$
　　$b_1 = 341 \cdot 10^6$ km
　　$p_1 = 251 \cdot 10^6$ km
　　$\tau_\pi = 2.732\,a$
(b) $r = 228 \cdot 10^6$ km
　　$\theta_1 = 81.23°$
(c) $v_1 = 29.63$ km/s
(d) $\dot{r}_1 = 15.30$ rad/s
　　$r\dot\theta_1 = 25.37$ rad/s
(e) $v_B = 24.12$ km/s
(f) $v_{en} = 15.42$ km/s
(g) $a^1 = 180$ km
　　$b^1 = 4174$ km
　　$\theta_\infty = 92.47°$

(h) $v = 4.93°$
(i) $v_{ex} = 15.42$ km/s
　　$v_2 = 30.68$ km/s
(j) $\Delta T = 15\,831$ GJ
(k) $\tau_{\mathrm{IS}} = 20.728$ h
(l) $a_2 = 596.3 \cdot 10^6$ km
　　$p_2 = 273 \cdot 10^6$ km
　　$\varepsilon_2 = 0.7363$
　　$b_2 = 403 \cdot 10^6$ km
　　$\theta_2 = 74.45°$
(m) $\alpha_2 - \alpha_1 = 1.5°$
(n) $r_{AP_2} = 1035 \cdot 10^6$ km
　　$\Delta r_{AP} = 257 \cdot 10^6$ km
(o) $1.8 \cdot 10^6$ km
(p) $1.71\,a$

10-6　　(a) $a = 19\,110$ km　　　　(b) $\varepsilon = 0.681$

10-7

(a) $\Delta v = -0.003\,468$ km/s　　(c) $\Delta v = -0.465\,263$ km/s
(b) $\Delta v = 0.286\,531$ km/s

10-8	(a) $\Delta v = -2.967$ km/s	(b) $a = 6850$ km

10-9	(a) $a = 2\,684.060 \cdot 10^6$ km	(f) $\theta = 80.2°$
	(b) $r_{PE} = 88.574 \cdot 10^6$ km	(g) $v_{en} = 51.23$ km/s
	(c) $r_{AP} = 4\,279.547 \cdot 10^6$ km	(h) $\delta = 0.87°$
	(d) $b = 683.835 \cdot 10^6$ km	(i) negligible

10-10	(a) $\Delta T = 8\,179$ GJ	(b) $b'_c = 3675$ km will not collide

Chapter 11

11-1	$\tau_4 = 2\pi \sqrt{\dfrac{a_1^3}{\mu'}} = \tau$

Chapter 12

12-1	$x_1 = 322\,070$ km
	$x_2 = 445\,114$ km
	$x_3 = -386\,226$ km

12-2	$r = 326\,740$ km

12-3	$\tau = 2.186$ ms/century

12-4	$C = 3.145$ km^2/s^2

12-5	(a) $C = 512$ km^2/s^2	(b) $778.3 \cdot 10^6$ km	(c) $v = 13.08$ km/s

Chapter 13

13-1	(a) $F = 23.854\ \mu N$	(b) $r_R = 65\,950$ km

13-2	(a) $\Delta S = 14.137$ m	(b) $t = 4$ min

13-3	$r_R = 143\,140$ km

Author Index

Subject Index